4

INSIDE RELATIVITY

Inside

Relativity

BY DELO E. MOOK
AND
THOMAS VARGISH

PRINCETON UNIVERSITY
PRESS

ACKNOWLEDGMENTS

"Leopards in the Temple." Reprinted by permission of Schocken Books Inc.
from *Parables and Paradoxes* by Franz Kafka, trans. by Ernst Kaiser and Eithne
Wilkins. Copyright © 1935 by Schocken Verlag, Berlin. Copyright 1936, 1937
by Heinr. Mercy Sohn, Prague. Copyright © 1946, 1947, 1948, 1953, 1954, 1958
by Schocken Books, Inc. Copyright © renewed 1975 by Schocken Books, Inc.

Our figure 1.1 is reproduced with permission from "The Curvature of Space in
a Finite Universe" by J. J. Callahan. Copyright © 1976 by Scientific American,
Inc. All rights reserved.

Our figure 6.2 are prints #2100, 2101, and 2300 from the catalog of the
National Optical Astronomy Observatories; use of these photographs does not
state or imply the endorsement by NOAO, or by any NOAO employee, of any
individual, philosophy, commercial product, process, or service.

Our figure C.1 is the first page from the article "Search for tachyon monopoles
in cosmic rays" by D. F. Bartlett, D. Soo, and M. G. White, *Physical Review D*,
volume 18, number 7 (1 October 1978); used with the permission of the American
Physical Society and D. F. Bartlett.

Our figure C.2 is reproduced by permission from *Nature*, volume 248, p. 28.
Copyright © 1974 Macmillan Journals Limited.

To   Kathryn, Richard, and Robert Mook

            and to

            Kathryn and Anthony Deering

In the temple of Science are many mansions.

—Albert Einstein, *Principles of Research*

Leopards break into the temple and drink to the dregs what is in the sacrificial pitchers; this is repeated over and over again; finally it can be calculated in advance, and it becomes a part of the ceremony.

—Franz Kafka, *Parables and Paradoxes*

CONTENTS

Preface

Einstein's theory of relativity is not easy to understand. Despite the plethora of attempts to explain it to "nonspecialists," including some by Einstein himself, most readers with little or no background in mathematics and physical science fail to find an accurate and comprehensible presentation. As the eminent historian of science Gerald Holton has put it:

> In fact, there is to this day no generally agreed source, the *reading* of which by itself will bring a large fraction of its nonscientific audiences to a sound enough understanding of these ideas, even for those who truly want to attain it and are willing to pay close attention.[1]

Inside Relativity attempts to fill this hiatus by means of a collaboration between two scholars, one of whom works in the physical sciences and the other in the humanities.

Our aim is to make the scientific theory of relativity vital and accessible. We assume that our readers have a serious interest in understanding the theory of relativity as developed in the physical sciences and are willing to accept the fact that modern physics can seem counterintuitive and paradoxical.

In fact, we think it is likely that our explanations will spark interest in more detailed aspects of the subject than we offer here, and we have at several points provided references to more elaborate discussions. These references are usually to studies suited to "general readers" with little scientific background, but there are also references for those who have studied some physical science. Occasionally we suggest works intended for serious students of physics, either because these contain a section that will be understandable to a general reader who has finished the present book, or because we were unable to suggest a reference at a more elementary level. Initial citations of references will always indicate their level of scientific sophistication.

We have tried to use language that is both clear and suggestive and at the same time faithful to the scientific ideas underlying relativity theory. We have permitted technical terminology to enter only after determining that it

[1] Gerald Holton and Yehuda Elkana, eds., *Albert Einstein: Historical and Cultural Perspectives* (Princeton, N.J.: Princeton University Press, 1982), p. xxv.

has genuine utility for a general reader. We have identified and defined each special term in a glossary.

This book requires no intimacy with mathematics. We do use one quantitative relation (distance traveled equals speed of travel multiplied by the time) at several points, but we trust that this will cause little difficulty. Equations appear very occasionally as *supplements* to the text, but they are always introduced with a full verbalization, and readers may choose to ignore them without losing the continuity of our exposition. Those willing to tolerate arguments based on graphs will find some points of special relativity theory elaborated in appendix A; appendix B supplements this discussion with algebra. Appendix C employs the results of appendix A to discuss features of hypothetical particles called "tachyons," particles that move faster than light. Appendix D discusses the meaning of Maxwell's equations, which were primary to Einstein's development of relativity theory. This appendix, independent of those preceding it, is also written for general readers with no special background in mathematics or physical science. These appendixes aim to satisfy readers who want a more complete justification for certain points that we offer in the main text, but are entirely supplementary.

Inside Relativity begins with an introductory chapter that serves a dual purpose: to suggest a framework for discussing scientific research, and to provide a general, broad, historical setting for the work of Albert Einstein. The next chapter establishes the immediate scientific context of relativity theory. In particular, we emphasize Newton's contributions to physics and we allude to the cultural influence of the Newtonian world view, the immense prestige accorded to Newton and his work down to the twentieth century. Chapter 3 is a discussion of Einstein's special theory of relativity based on his original paper of 1905. We have included an explanation of how the theory was recast into the framework of what is called "four-dimensional spacetime," and what this reformulation means. The next chapter discusses some consequences of the special theory, including some of the famous so-called "paradoxes" that result from its application. We offer a discussion of the various effects that cause visual "distortion" of objects observed in motion at speeds approaching that of light. We have calculated the appearance of a variety of three-dimensional objects moving by at various speeds and viewed from various distances, and these calculations are illustrated in the figures. Chapter 5 presents Einstein's general theory of relativity and some of the conclusions to be drawn from it. Again our explanation is based, initially at least, on one of Einstein's first summary papers on the subject. Chapter 6 discusses the cosmological applications of general relativity in some detail; this discussion contrasts the general relativistic view of the cosmos with the pre-Einsteinian views developed in chapters 1 and 2 so that the impact of Einstein's work on our conceptions of the universe is made clear.

We have been liberal in our use of illustrations. We believe that a combination of carefully chosen verbalizations with simple graphical illustrations is the most effective way to communicate physical concepts to nonscientists. At the same time we recognize that some of our readers, accustomed to entirely verbal expositions of ideas, may at first find the figures a worrisome "interruption" to the flow of the discussion. Nevertheless, the figures are thoroughly integrated with the text and are meant to be *read* along with it. A few moments spent pondering a figure associated with a passage that appears difficult will well repay the effort. Sometimes the graphical "exposition" is built up through a series of similar diagrams, each evolving from the one preceding it. This style of presentation permits rather sophisticated diagrams to be developed and used without risking the intimidation that can result when a single and very complicated or visually "full" figure is used. Certain figures in the text are referred to parenthetically as "optional." These convey exceptionally detailed information that is not essential for understanding our arguments but that may be of interest. On the other hand, if certain of the figures seem trivial we urge our readers to remember that in relativity theory the simplest motion engages us in the largest questions of temporal measurement, and that the illustration of a ship's deck moving on the ocean or a flatcar on a few meters of railroad track can invoke the observation of galaxies in their courses.

Finally, a note about the way we make cross-references to material within the book. Each chapter is divided into numbered sections. When we refer to one of these sections it will appear as a chapter and section number. For example, 2.4 refers to section 4 of chapter 2.

WE KNOW that our approach to describing relativity theory works because we have tested much of it on audiences similar to those for whom this book is written. These were participants and faculty in sessions of various executive development seminars such as those sponsored by IBM, Control Data Corporation, United States Steel Corporation, the United States Treasury Department, and most especially by the Dartmouth Institute, a four-week program in liberal arts for professional and business people and their spouses. The questions and criticisms offered during these sessions by our fellow faculty and by other participants helped us to understand how relativity theory can be so central to our modern intellectual identity and yet so difficult to communicate. We would like to take this opportunity to thank Dartmouth Institute participants and our colleagues on the faculty. We wish to acknowledge especially the help provided by Professors Agnar Pytte, Stephen Nichols, Ronald Green, James Wright, Mary Kelley, Charles Drake, James Hornig, and Director of Continuing Education, Gilbert Tanis. Alice Calaprice, our editor at Princeton University Press, has provided sustained creative assistance in preparing the manuscript for publication. Sylvia Scherr and Anthony Deering read preliminary drafts most thoughtfully and provided many valuable comments and suggestions. Nancy Bronder also read the manuscript and through her enthusiasm for our project and stimulating conversations contributed to its progress in many ways. We are grateful to Elizabeth Deeds Ermarth for her support and advice throughout the writing of this book.

INSIDE RELATIVITY

1

THE SCIENTIFIC ENTERPRISE: POSING MODELS

> Only in men's imagination does every truth find an effective and undeniable existence.
>
> —Joseph Conrad

1.1 INTRODUCTION

This book is written to explain Einstein's theories of relativity to nonscientists; it treats the content and the development of Einstein's work. Many of the ideas that we will be discussing (particularly those involved in the special theory of relativity in chapters 3 and 4) are intrinsically quite simple. But certain conclusions drawn from these ideas have proved persistently counterintuitive—so much so that readers encountering the theory are often left with a feeling of bewilderment or even of intellectual loss. The reasons for this sense of bewilderment and loss have to do with the nature of the scientific revolution resulting from Einstein's work and with its implications for nonscientific aspects of our culture. We suggest that this revolution is of a different order from the changes wrought by the work of Darwin, or of Newton, or of Copernicus and Kepler. Those great developments were understood with a modest effort by intellectuals of the periods in question because the changes could be accommodated by habits of mind and modes of thought already in existence. Unlike the contemporaries of Darwin and Newton, as we proceed with our discussion of Einstein's work we will be challenged to alter some of the fundamental organizing principles and premises upon which we base our perceptions of the world.

This opening chapter establishes some necessary background. By viewing Einstein's work in the context of its scientific antecedents, we can appreciate just what Einstein did as a scientist and how his creations have challenged our understanding of the physical world. In order to open this larger view we will suggest a general paradigm for the methods of the scientific enterprise. Our remarks about the operation and structure of science will be elaborated through a series of related examples drawn from its history. We have chosen these not only because they illustrate our generalizations about the nature of science, but also because they establish key reference points that bear directly on the scientific and cultural context of Einstein's theories of relativity.

1.2 MODELS: THE PRODUCTS OF SCIENCE

Certain summarizing results of scientific research are frequently described by such terms as "theory," "hypothesis," or "law of nature." One also hears about something called "the scientific method" or "the scientific approach," which can suggest that there is some well-defined procedure for making scientific discoveries or arriving at the results of scientific research. Some of the confusion that nonscientists have in reading about science comes from the attempt to fit a discussion of actual scientific research into a preconceived notion of a standard "scientific method." As with any other creative effort, scientific research may follow a variety of disparate patterns, and sometimes the process of science will seem enigmatic, defying any attempt to discern a pattern or method (the "stroke of genius" or "brainstorm" come to mind here). It is probably best to put aside ideas of a set "scientific method" in examining the creative acts of scientists, but this is not to say that science as an endeavor lacks structure. As we will explain presently, that structure is imposed not so much on the creative acts of the scientist as on the methods by which scientific ideas are tested and valorized and by the need for integrity and logical consistency within and among scientific theories.

In this chapter we prefer to use the term "model" to describe the products of the scientific enterprise. Scientists themselves frequently use this word, and we adopt it because it suggests explicitly that the final results of scientific inquiry are "free inventions of the human mind,"[1] to use Albert Einstein's description of physical concepts. These free inventions are meant to represent some aspect of the perceived physical world, and in this respect science resembles much artistic endeavor. In its attempt to represent the world science proceeds with its fair portion of the folly and bias

---

[1] Albert Einstein, *Essays in Science* (New York: Philosophical Library, 1934), pp. 15 and 16; Albert Einstein and Leopold Infeld, *The Evolution of Physics* (New York: Simon and Schuster, Touchstone Books, 1966), p. 31.

that go with any human activity. On the other hand, it would not be accurate to say that science is exactly like the arts, for scientific models may not be validated in quite the same manner as works of art. Scientific models are regarded as more or less *useful* for describing, understanding, and manipulating the world and, as we will see, it is largely this perceived *utility* that determines the value and the prestige accorded a model.

1.3 THE GEOCENTRIC MODEL

To illustrate the process by which the prestige of models waxes and wanes, we have chosen to discuss some cosmological models created to comprehend the observed motions of the sun, moon, and planets. The first one we consider is found far back in antiquity. We call it the "geocentric" model, although the word "Ptolemaic" is sometimes used instead. "Ptolemaic" refers to the great astronomer Claudius Ptolemy, who refined the geocentric model extensively and whose treatise, *The Almagest*,[2] which explained his system, became the basis for medieval astronomy.

Here is one verbal statement of this model:

> *The sun, moon, planets, and stars all move about the earth,*
> *which is at rest in the center of the universe.*

Geocentric models of the cosmos have been used longer than any other kind, and they continue to serve us well in our daily activities. For example, we still use the words "sunrise" and "sunset," which are derived from this model and which imply that the sun moves around our stationary earth. Our everyday experience appears to suggest that the earth is indeed stationary, and that if something seems to move with respect to the earth, then that object, and not the earth, is doing the moving.

A pictorial statement of a geocentric model is shown in figure 1.1. This illustration is from Dante's *Divine Comedy* (originally published about 1320).[3] While the scientific model it conveys is in essence that given in the verbal statement, the picture adds significant detail, some of it of a "non-scientific" nature if judged by present scientific criteria. In the figure, we

---

[2] A recent English translation is that by G. J. Toomer, *Ptolemy's Almagest* (New York: Springer-Verlag, 1984).

[3] Copied from J. J. Callahan, "The Curvature of Space in a Finite Universe" in *Cosmology Plus One*, ed. Owen Gingerich (San Francisco: W. H. Freeman and Company, 1977), p. 30. This book is a collection of articles from *Scientific American*. General readers will find these articles challenging but worthwhile. More extensive references to this book will be made in chapter 6, which deals with cosmology.

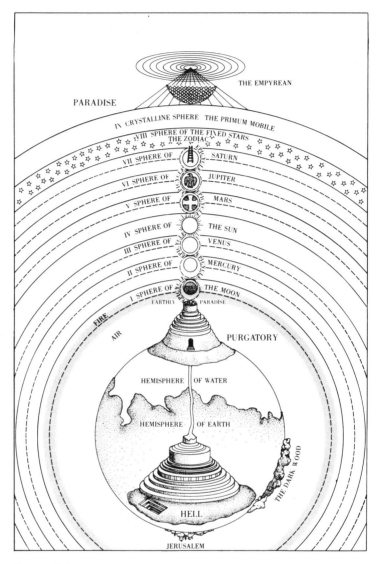

FIGURE 1.1

see a schematic representation of the various celestial objects that are supposed to be fixed to transparent spheres, each turning about the central
earth. The ninth sphere contains the *primum mobile* or "prime mover,"
which turns perpetually and in so doing governs the motions of the other
celestial objects. But there is more. This statement of a cosmic model is

also identified with what we would distinguish as theological doctrine. The theological elements, which can appear strange to us today, are established through certain features of the physics that complements this model—largely the physics of Aristotle, some terms of which will be familiar to many of our readers.

For medieval thinkers, matter in the sublunary sphere (the region beneath the sphere containing the moon), including all matter on earth, was composed of the four elements—fire, air, earth, and water—in various combinations. Each of the elements had specific properties that could be used to explain a variety of phenomena. For example, one of the properties of the element earth is that it always seeks the center of the universe (assumed to be located at the center of the spherical earth). The element fire, on the other hand, always seeks heaven, and so it tends to rise up into the sky. Objects that contain more of the element earth will therefore exercise a stronger attempt to reach the center of the universe and so will be heavier; objects containing more of the element fire will try to rise more vigorously, and so they are lighter. People influenced by modern science may find these explanations of (what we now call) "weight" quaint rather than powerful. But to medieval scholars the models of Aristotelian physics represented a highly serviceable way of organizing and interpreting their observations of the world.

Beyond the sublunary sphere, objects were made of a special fifth element (the quintessential element) called "ether," an element not found on the earth or anywhere within the sublunary sphere. Anything existing above the sublunary sphere was made of ether and was supposed to be changeless, in poignant contrast to the way things were observed to be here on earth. Unlike the four terrestrial elements, ether had the property of perpetual motion—following the perfect geometric curve that has no beginning or end, the circle. Therefore, the sun, moon, planets, and stars, because they were made of ether, were thought to follow perpetual circular motions in space. These natural circular motions of ether were, in turn, used by astronomers to explain observed celestial motions.[4]

The explanation for the great difference between the decaying and imperfect sublunary sphere and the perfect volume beyond the sphere of the moon had to do with distance from the outermost region of the cosmos, the Empyrean, the abode of God. The nearer something is to God, the greater its degree of perfection and the greater its state of grace; the farther from

---

[4] We will see in chapter 2 that the term "ether" was used by physicists in the nineteenth century in a very different context.

God, the more base and corrupt it must be. The orbit in which the moon moved marked the dividing line between the region of celestial perfection and the region of the universe containing all corruption.

The view of the cosmos illustrated in figure 1.1 shows these notions carried explicitly into the spiritual world. Hell is inside the earth, even farther from God than our immediate terrestrial surroundings. Hell itself is divided into a number of concentric levels, each farther from God than the last. Sinners are consigned to one of the circles of hell according to the magnitude of their sins: the greater the sin, the deeper and the more horrible the circle of punishment—that is, the farther one is sent from God. And so the geocentric model, created to explain celestial phenomena, was also applied in contexts a modern scientist would claim are far removed from observations of astronomical phenomena. The hierarchy illustrated in figure 1.1 was reflected in the theological (and even the political) structures of the Middle Ages.[5] The perspective we gain by considering a model like this, a now "quaint" model that once enjoyed high prestige, emphasizes the fact that conceptions of the physical world are liable to be freighted with value. The creation and the valorization of models takes place under the full burden of the scientist's cultural background. That background, in turn, is shaped in part by the interpretations given to experiences of the physical world.

So far we have given verbal and pictorial statements of the geocentric model. We also can use the language of mathematics to represent it by drawing geometric figures or by writing equations. In such a representation, mathematical symbols are analogous to the verbal or graphic symbols of the other statements. But whether the medium of expression is verbal, graphic, or mathematical, a geocentric cosmos is represented. The different expressions of the same underlying physical concept may be more or less sufficient, depending upon the use to which the model is to be put, and these different expressions can have differing heuristic value. For example, a mathematical statement does not in our day appear to lend itself to the inclusion of spiritual or moral elements. On the other hand, a mathematical representation of the geocentric model does permit detailed, quantitative predictions for the positions of celestial objects. These predictions constitute a practical aid to astronomers and navigators in forecasting celestial events, and they can also be compared with actual observations of celestial objects to check the accuracy of the model's predictive power. This provides what is usually considered to be an "objective criterion" for evaluating the utility of the model.

---

[5] See, for example, the discussion by Owen Barfield, *Saving the Appearances* (New York: Harcourt, Brace & World, 1965), especially chapters 11–13.

1.4 THE HELIOCENTRIC MODEL

As the precision of astronomical measurements grew, the failure of existing geocentric models to predict the future positions of celestial objects with sufficient accuracy became obvious and another sort of model, the heliocentric model, gained prestige. Here is a simplified verbal statement of a heliocentric model:

The planets, including the earth, move around the sun.

Descriptions of heliocentric models can be traced back to the time of Herakleides (388–315 B.C.) and Aristarchus (310–230 B.C.), but the key personality in the establishment of this model's dominance some 1800 years later was Nicolaus Copernicus (1473–1543). The uproar caused by the overthrow of the geocentric model has since been called "the Copernican Revolution," although this phrase is probably inexact.[6]

Actually, the version of the heliocentric model taught in schools today is a highly evolved version of the Copernican idea, created largely by Johannes Kepler (1571–1630). Kepler published three rules, or "laws," of planetary motion. These statements describe the shape of planetary orbits about the sun (the orbits are slightly elliptical), the manner in which the speed of each planet changes in its orbit, and the relationship between the average distance of the planet from the sun and the amount of time the planet requires to complete one orbit. Kepler's laws represented a significant advance in simplicity and in accuracy of prediction (utility) for the heliocentric model, and so they supplanted Copernicus's model.

All of us are familiar with the heliocentric model. It has been taught to us since childhood. Yet our sense impressions do suggest that the earth is at rest beneath our feet and not moving through space around the sun; our language continues to use terms characterizing the geocentric model ("sunrise" and "sunset"); and in navigation and many sorts of timekeeping the geocentric model is still employed. Nevertheless, we have been taught that a heliocentric model is the acceptable type to use in describing "astronomical reality." Some may even have become convinced that in some sense Kepler's heliocentric model is a "true" model, the "best" model, or even

---

[6] The revolution required more than Copernicus's contribution: the work of Kepler was especially important. See Thomas S. Kuhn, *The Copernican Revolution: Planetary Astronomy in the Development of Western Thought* (Cambridge, Mass.: Harvard University Press, 1957). This is an excellent book by an eminent historian and philosopher of science written so that it is easily accessible to the general reader. See also the more recent discussion in I. Bernard Cohen, *Revolution in Science* (Cambridge, Mass.: The Belknap Press of Harvard University Press, 1985), chapters 7 and 8.

"the" true model. But it is none of these, as we will soon see. It is also worth bearing in mind that while a heliocentric model may be the type granted the greatest prestige in schools today, it was definitely not the favored model when it was published by Copernicus; it posed very serious problems for its advocates, particularly for Galileo Galilei (1564–1642).

Galileo's troubles, culminating in his notorious trial and conviction by the Inquisition, are worthy of our careful attention because in this episode we find an explicit contrast between the process of model validation that many of us today accept as "self-evident" with another process that played a significant part in our cultural history. The heliocentric model challenged nonastronomical foundations of medieval physical science. Its acceptance required rebuilding a substantial portion of the intellectual framework then used for understanding the world. We know that for pre-Copernicans the nature of the four elements from which matter is made and the way objects move are intimately related to the structure of the astronomical cosmos. The earth is at the center of the universe, and the natural motions of the various elements are related to that central position. In particular, the elements fire and earth move in a manner prescribed by the central, stationary position of the planet earth. In any heliocentric model the earth is no longer in a state of rest at the center, and so the reference frame necessary for making sense of the properties and motions of the elements (and of objects made from the elements) is destroyed.

In addition, acceptance of a heliocentric model of the cosmos required that the earth be just one of several planets orbiting the sun. According to medieval physics, this is impossible because only the perfect, celestial bodies made of ether can orbit perpetually in space. If the earth does in fact orbit, then perhaps the other planets—even those closer to the Empyrean than earth—are not made of ether and may be as imperfect as the earth itself. But how can the heavens so close to God be imperfect when since antiquity it has been accepted that they are perfect and changeless? Acceptance of a heliocentric model was tantamount to a destruction of the way people looked at matter and the way it moves. It robbed them of the understood properties of the elements that make up the world.

There was still another practical problem based on rather elementary arguments and an appeal to "common sense" (in principle not unlike some "common sense" appeals raised in opposition to the results of Einstein's relativity theory some three hundred years later).[7] If a heliocentric model is

---

[7] See, for example, the discussion in Cohen, *Revolution in Science*, pp. 413–414 and 376–377.

correct, then once each year the earth must travel all the way around the sun in space along a huge path. The size of that path could be estimated, and so the speed required for the earth to cover that path in one year could be calculated. That speed turns out to be about 67,000 miles per hour. If one believes in a heliocentric model, one must also believe that at each instant the earth is rushing through space at about 67,000 miles per hour. If this were true, how could people possibly stay on the rapidly moving earth? Why isn't everyone left behind in space? And wouldn't we "know" (that is, receive appropriate sense impressions) if we were moving at such an outrageous speed all of the time? Furthermore, because heliocentric models such as Copernicus's and Kepler's require that the moon move around the earth while the earth rushes around the sun, a Copernican must also explain why the fleeting earth does not simply leave the moon behind in space.

These problems were difficult for the early followers of Copernicus; they could devise no convincing responses to them because the available models of nature were inadequate. In other words, justification for the heliocentric vision of the cosmos could not be found in the models of physics current at that time. This point is thoroughly discussed by I. Bernard Cohen in his book *The Birth of a New Physics*, where the "new physics" is the one demanded by the acceptance of the heliocentric system.[8]

There was even a political problem with the heliocentric system. It had to do with the way in which moral and spiritual doctrine complemented the geocentric system. That system was defined with reference to the spiritual order, and the dogma of the Church specified that the earth is at rest in the center of the cosmos. This idea was, of course, also consistent with the authority of Scripture. Consider Psalm 93:

> The Lord reigneth, he is clothed with majesty;
> The Lord is clothed with strength wherewith he hath girded himself;
> The world is stablished, that it cannot be moved.

It is comforting to have the cosmos arranged in a very orderly way about us on a central earth. Human beings, created in the image of God, are then at least near the center, even if they are in the most base part of the whole cosmos. For those accustomed to feeling at the very focus of cosmic concern, the overthrow of the spiritual cosmos demanded by a heliocentric

---

[8] I. Bernard Cohen, *The Birth of a New Physics* (New York: W. W. Norton & Company, 1985). This is an excellent history of the models discussed in this chapter and is written in language appropriate for the general reader.

model was not simply a matter of what we would now call "scientific" understanding; it contained a profound and personal threat.

Despite these and other difficulties involved in accepting a heliocentric model, the so-called Copernican revolution began. Copernicus's (and later Kepler's) system had certain conceptual advantages over the geocentric system in explaining the observed motions of planets. Usually, for example, the position of the planets in the sky shows a gradual eastward shift with respect to the stars. But on occasion each planet will move westward against the background of stars for a few days (this is called "retrograde motion"). The geocentric model of Ptolemy postulated (in a completely ad hoc way) special devices to account for this motion. Heliocentric models can account for this phenomenon as a simple geometric consequence of the fact that the planets and our terrestrial observing platform are all in orbit about the central sun. Heliocentric models also account for the fact that these retrogradations of the planets occur only when the sun and the planets have a special alignment in the sky, something that geocentric models, again, had to explain for each case. And so a number of thinkers began to seek solutions to the problems raised by the heliocentric model and to provide evidence for its correctness.

1.5 GALILEO

In 1609 Galileo Galilei built a small telescope and pointed it toward the sky. Galileo was almost certainly not the inventor of the telescope (as is sometimes stated) and he may not even have been the first person to observe the sky with it, but he is honored because he was clearly the first to study celestial phenomena carefully with this revolutionary technology and to use his observations to solve important scientific problems. He published detailed accounts of his work rapidly, sometimes even in vernacular Italian rather than in scholarly Latin so that the general public could hear of his remarkable discoveries.[9]

We will cite three of Galileo's telescopic discoveries. Each can be verified and understood by anyone caring to look at the sky through a telescope, and each turns out to be strong empirical evidence against geocentric models and for a heliocentric model.

---

[9] Many of his telescopic observations are described in his book, *Sidereus Nuncius*, published shortly after he made his telescope. See Galileo Galilei, *The Sidereal Messenger*, trans. Edward Stafford Carlos (London: Dawsons of Pall Mall, 1964). Galileo's publications make enjoyable reading for scientists and nonscientists.

Imperfections in the Heavens

According to medieval dogma, the moon was a perfect celestial body, and yet Galileo observed that it was blemished with mountains and hills and valleys. He was even able to measure the altitude of some of the lunar features and to show that their elevations could exceed the highest known on earth. Clearly, the moon was no perfect celestial sphere. Galileo also observed the sun on several occasions and noted the phenomenon of sunspots. Even the sun, much closer to God than the moon or earth in a geocentric system, has blemishes on its surface, imperfections that again deny the supposedly perfect nature of celestial objects. Galileo also discovered that the pattern of sunspots undergoes changes in time, thus giving evidence of celestial mutability as well.

Mountains on the moon or changing spots on the sun may seem harmless enough from our modern perspective, but at the time of their discovery they demonstrated strong empirical evidence (a kind of evidence that was developing prestige) for imperfection and change beyond the sublunary sphere. The properties of the ether were not what thinkers had supposed in the past, if indeed the ether were the substance of celestial objects.

The Jovian Moons

We have mentioned that one argument against the heliocentric system was based on the belief that the earth could not possibly orbit the sun while the moon orbits the moving earth. The high speed of the earth in space, it was argued, would cause the earth to leave the moon behind. When Galileo turned his telescope to the planet Jupiter he made another remarkable discovery that tended to refute this argument: four "new stars" that moved about the planet in regular cycles. In other words, Galileo had discovered that Jupiter is attended by four moons. Given its distance from earth, in both the geocentric and heliocentric models, Jupiter must be moving through space at a very high speed in order to account for its observed rate of motion against the stars in the sky. Galileo had found that the swiftly moving Jupiter is attended by not just one moon (as is the earth in heliocentric models such as those of Copernicus or Kepler) but by four moons that are not left behind by Jupiter's swift motion. This gave clear evidence (again of an empirical sort) that there was something very wrong with the physics of Galileo's day. Whether or not a heliocentric model was accepted, medieval physical science was unable to explain the existence of the Jovian moons.

The Phases of Venus

When Galileo observed the planet Venus through his telescope he found
that it demonstrated phases like the moon. The phases changed over the
course of months from nearly full to nearly new. This observation represents a phenomenon that the geocentric system cannot explain. Anyone can
understand why, whether or not they are trained in mathematics or astronomy, and so this observation was readily recognized to be a powerful (empirical) refutation of the Ptolemaic model.

To appreciate the refutation, it is necessary to understand some of the
special properties of the planet Venus and the way a geocentric system had

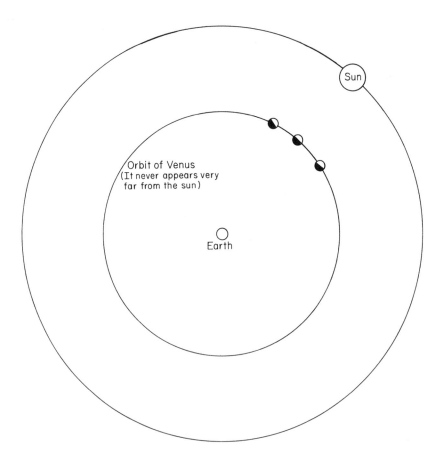

FIGURE 1.2

to explain them. First, Venus is never observed to be very far from the sun in earth's sky. Figure 1.2 shows a schematic diagram of the geocentric model of Venus's orbit about the central earth. The geocentric system is made consistent with the observed proximity of Venus and the sun by supposing that they move more or less together about the central earth. There is no a priori reason for this synchronous motion of the sun and Venus. The synchronization was added in an ad hoc manner to account for the observed close proximity.

Because the sun always appears close to Venus, on earth we see mostly the ''back'' or unilluminated side of the planet. Therefore, Venus should show a dark (or ''new'') disk or at most a thin crescent phase when seen from earth. This fact had been recognized since antiquity, but the disk of Venus is so small in the sky that to the unaided eye it appears as a point of light. Galileo's telescope, however, revealed the disk of Venus and its phases for the first time. The observed phases included thin crescents as well as all phases from nearly new to nearly fully illuminated disks.

Figure 1.3 is a schematic diagram of the heliocentric model of Venus's orbit about the central sun. Now the proximity of Venus to the sun in earth's sky receives an inherent geometric explanation without an ad hoc synchronization of motions. Venus orbits the sun at a smaller distance than does the earth; therefore, Venus can never appear to be very far from the sun when seen from earth. In this model, when Venus lies between the sun and the earth, we will see a thin crescent; when it lies on the far side of the sun from earth, we will see a more fully illuminated disk. Furthermore, notice that the more fully illuminated the disk, the farther Venus is from earth and so the smaller it should appear in the sky. This was exactly as Galileo observed with his telescope. The appearance of Venus in the telescope contradicted the geocentric model of Ptolemy and argued for a heliocentric one.[10]

Galileo accumulated this and other evidence against the geocentric model of Ptolemy and publicly advocated the heliocentric model of Copernicus. But, as we have seen, to advocate a heliocentric model was to advocate the rejection of some of the basic tenets of contemporary science and to undermine the theological system of the day. It was also to take a

[10] This observation did not unequivocally *prove* a heliocentric model, however, because it turns out that a heliocentric model is not unique in its ability to explain the phases of Venus. On the other hand, the heliocentric model was the alternative having the greatest prestige in Galileo's day and for three centuries subsequently. See Kuhn, *The Copernican Revolution*, pp. 222–225, for a discussion of this point.

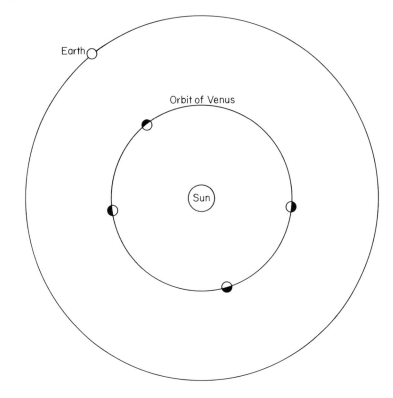

Earth

Orbit of Venus

Sun

FIGURE 1.3

stand in direct opposition to church doctrine. As a result of his advocacy, and most immediately because of the publication of his book *Discorsi E Dimostraziono Matematiche, Intorno A Due Nuoue Scienze* (Dialogues Concerning Two New Sciences),[11] Galileo, at the age of seventy, was brought to trial for heresy before the Inquisition. There is no evidence that he was subjected to physical torture by the Church, but in the "Book of Decrees of the Congregation of the Inquisition" of 1633 it is recorded that Galileo was "to be interrogated . . . even threatened with torture, and if he sustains it, proceeding to an abjuration."[12]

Galileo signed the abjuration. The nature of Galileo's crime is clear from

[11] An English translation is Galileo Galilei, *Dialogues Concerning Two New Sciences*, trans. Henry Crew and Alfonso de Salvio (New York: Dover Publications, Inc., 1954).

[12] This quotation is taken from the article by Owen Gingerich, "The Galileo Affair," *Scientific American* 247 (August 1982): 142.

this document: he believed in a heliocentric model including the moving earth and he refused to submit to the authority of the Church by abjuring his beliefs. Galileo was sentenced to house arrest in 1633; he died nine years later at age seventy-eight, blind, and still under detention.

From our modern perspective, the notion of a moving earth is so familiar that it can be difficult to understand why the Church was so exercised about it; but as we have tried to show, a great deal more was at stake in Galileo's case than the issue of the earth's motion. The very *idea* of the moving earth was considered by some too dangerous not to suppress. The heliocentric model threatened to rob people in the seventeenth century of their familiar way of thinking about the world and of expressing feelings about their surroundings. Consider this famous expression of anguish by John Donne, a contemporary of Galileo's, who bemoans the acceptance of the heliocentric model:

> And new Philosophy calls all in doubt,
> The Element of fire is quite put out;
> The Sun is lost and th'earth and no man's wit
> Can well direct him where to looke for it.
> And freely men confess that this world's spent,
> When in the Planets, and the Firmament
> They seeke so many new; they see that this
> Is crumbled out againe to his Atomies.
> 'Tis all in peeces, all cohaerence gone;
> All just supply, and all Relation.[13]

The Galileo affair obviously represents more than a conflict between a threatening model and the status quo, as his signed abjuration makes clear: "After an injunction had been judicially intimated to me by this Holy Office to the effect that I must altogether abandon the false opinion that the Sun is the center of the world . . . and that I must not hold, defend, or teach in any way whatsoever . . . the said false doctrine, and after it had been notified to me that the said doctrine was contrary to Holy Scripture—I wrote and printed a book in which I discuss this new doctrine."[14] He had refused, in other words, to submit to the authority of the Church. This represents a conflict of values brought to a focus in what amounts to an epis-

---

[13] John Donne, "The First Anniversary," in *The Poems of John Donne*, ed. Herbert J. C. Grierson (Oxford: Oxford University Press, 1912), I, 237.

[14] This translation is taken from Arthur Koestler, *The Sleepwalkers* (New York: The Universal Library, Grosset & Dunlap, 1959), note 43 on pp. 607–608.

temological dispute. There is, first of all, the question, How should any model about the physical world be validated? But there is more to it than this suggests. Galileo was not just posing a scientific model in the sense that we have defined them: he was also asserting that his model was *a truth* about the celestial phenomena. So the second question raised is this: Who determines what is *true* concerning natural phenomena? Our distance from the inquisitors of Galileo and perhaps from Galileo himself is not measured so much in terms of a difference in accepted models of nature, but in the philosophical difference between those who are willing to live with a plurality of possible models and those who insist on just one vision of reality, "the truth," mandated by an authority.

Despite the strong empirical evidence that Galileo gathered in support of the heliocentric and against the geocentric model, he was unable definitively to resolve the debate over which was the better astronomical system. He could not explain, even to those who acknowledged empirical validation, what was wrong with the physics of his day. Why, indeed, *does* the moving earth not leave the moon far behind in space? Or, how can one account for the seemingly perpetual motion of the earth in a heliocentric system?

1.6 A NEW PHYSICS: ISAAC NEWTON

Galileo died in 1642; in the same year, on Christmas Day,[15] a person was born who would supply the physics necessary to understand the heliocentric model. This was Isaac Newton (1642–1727). By the age of thirty Newton had invented the calculus, the reflecting telescope, and the foundations for his monumental theory of mechanics—the name given to the branch of physics that deals with the way objects move and respond to the action of forces. We will discuss Newton's work in detail in chapter 2, for it was Newton's mechanics that provided the basis for understanding the heliocentric model of the cosmos as well as the basis for all of physics up to the early years of the twentieth century. It was Newton's mechanics that Albert Einstein's work in relativity would supplant. "Newton, forgive me," Einstein wrote in his "Autobiographical Notes," "you found the only way which, in your age, was just about possible for a man of highest thought—and creative power. The concepts, which you created, are even today still guiding our thinking in physics, although we now know that they will have

---

[15] Actually the date was January 4, 1643, according to the calendar now in use.

to be replaced by others farther removed from the sphere of immediate experience, if we aim at a profounder understanding of relationships.''[16]

In 1687 Newton published his *Philosophiae Naturalis Principia Mathematica*, the most famous work in physics and one of the most important books in the history of western thought. In three simple statements in the first few pages of this work, Newton stated his new model for explaining and predicting how objects move. He went on later in the book to describe what was then a new sort of force in nature, the force of gravity which, Newton supposed, acts between all masses in the universe and tends to draw them together.

When Newton combined his mathematical model of the gravitational force with his three statements describing the motion of objects, he found that he was able to predict all of the features of Kepler's heliocentric model of planetary motion and to explain, for example, why the moon is not left behind the rapidly moving earth. The moon is not left behind (and we do not fall off) the moving earth because the force of gravity draws things to the earth, just as the sun pulls on the earth in its own orbit in space. Newton's explanation for the perpetual motion of the earth will be discussed in detail in chapter 2; our point here is simply that Newton provided the tools to analyze, to understand, and accurately to predict the phenomena of the natural world here on earth or far out in space. And the prestige that accrued to Newton's models by virtue of their success deeply influenced our culture. In fact, it is fair to say that we are largely Newtonians in our outlook on the world. Hence Einstein felt the need to apologize as he began to undermine the prestige of what has been called ''the Newtonian world view'' that dominated physics for more than two hundred years. It largely constitutes the immediate scientific background for Einstein's work. One cannot appreciate the significance or origins of relativity theory without first understanding how it arose in a conflict involving Newtonian physics, and how Newton's models were ultimately superseded.

1.7 BEYOND NEWTONIAN PHYSICS

To prepare the way for our detailed discussion of the Newtonian world view and Einstein's relativity theory, we will briefly describe models of celestial phenomena that continued to evolve after Newton's time. When

---

[16] Albert Einstein, ''Autobiographical Notes,'' in *Albert Einstein: Philosopher-Scientist*, ed. Paul Arthur Schilpp (New York: Harper Torchbooks/Science Library, Harper & Brothers Publishers, 1951), I, 31–33.

Newton's model is compared with sufficiently precise observations of celestial bodies, significant discrepancies are found; in other words, when extreme accuracy is required the Newtonian model ceases to be useful. Probably the most significant of these failures of Newtonian theory for the
history of Einstein's relativity theory has to do with a discrepancy between
the predicted and observed motions of Mercury, the planet closest to the
sun. Mercury's orbit is found to be elliptical (as stated in Kepler's model
and as predicted by Newton's models), and it is closest to the sun at the
"perihelion" point of its orbit (figure 1.4). The location of the perihelion
is not fixed in space; it moves about the sun in what is termed "the advance
of the perihelion" because the orientation of Mercury's entire elliptical orbit rotates in space once around in a little over 23,100 years. Newton's
models could explain 99.2 percent of the observed perihelion advance of
Mercury as a result of the combined gravitational influences of the other
planets (superimposed on the dominant gravitational attraction of the
nearby sun). But by 1860 physicists and astronomers realized that despite
strenuous efforts, Newton's model could not predict the remaining 0.8 percent of the observed perihelion advance, an appreciable discrepancy of
about 185 years. Like all previous models of planetary motion, when
pressed too hard Newton's model was found to be inaccurate.

But there is another model, more accurate than Newton's, which we can
call upon to help with the problem of predicting the positions of planets
(and in particular with the advance of Mercury's perihelion) when New-

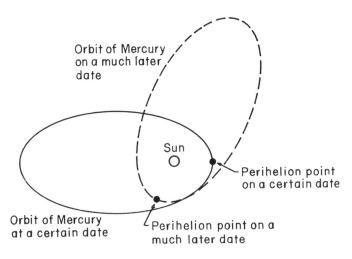

Orbit of Mercury
on a much later
date

Sun

Perihelion point
on a certain date

Orbit of Mercury
at a certain date

Perihelion point on a
much later date

FIGURE 1.4

ton's model no longer suffices. This is the model called "general relativity theory," and it was published by Albert Einstein in 1915. We will discuss it in detail in chapter 5. Here we hasten to add that even Einstein's model has its limitations. If it is pressed too far in certain circumstances, it will not provide the desired answers. Unfortunately, when general relativity theory does not serve, we have no better model to use, at least at present. Yet we do recognize that even the most accurate model available has its limitations, and that to satisfy more stringent demands of predictive capability a newer model must still be created.

1.8 SUMMARY: THE CHARACTERISTICS OF SCIENTIFIC MODELS

The cosmological models of the preceding sections have been discussed in some detail because they serve to illustrate the scientific enterprise as a creative effort. In particular, we are able to list six features of scientific models that are illustrated by the examples just given.

First, there seems to be an identifiable idea behind a model that can be expressed in a number of ways. We have seen that pictures, words, or mathematical formulae can be used to communicate the essence of the heliocentric or the geocentric models. Each mode of expression will have certain advantages, depending upon the use to which the model will be put. For example, to help in understanding the structure of the cosmos, a verbal or a pictorial statement of a model might be most useful, while a mathematical statement would be better for predicting the position of a planet on a given date.

Next, we find that all scientific models have limited domains of validity. When pushed beyond a certain point, models will not accurately describe what we observe in the world. In fact, scientific research is often directed toward determining the limitations of a model, as opposed to the creation of a wholly new one. The limitation in the domain of validity might be one of computational accuracy. For example, the Ptolemaic model will successfully predict the future positions of celestial objects if one does not demand too much precision. The model will not work if very precise predictions are required. On the other hand, the Ptolemaic model will not explain the observed phases of Venus. That application exceeds the domain of validity of the model, not because of computational accuracy, but simply because the model is an inappropriate one to use in dealing with the phenomenon of Venus's phases; planetary phases fall outside the domain of validity of the Ptolemaic model.

Third, scientific models change or evolve with time. This corresponds to the fact that they are transitory in usefulness and prestige. The change is sometimes characterized by a minor modification of a preexisting model to widen its domain of validity; sometimes a model is substantially altered or even completely replaced. In any case, new leopards are continually entering the temple, and the scientific enterprise is characterized by the continual creation of models to cover an ever-widening domain of validity, where that domain might include more phenomena or increase the predictive accuracy of the model. For example, we have discussed how, with time, new models of the cosmos were created to describe planetary positions with greater accuracy. In general, the models used and taught most widely today are not those used in the remote past, and they probably will not continue to be used indefinitely into the future. But we do sometimes live with a plurality of working models in that older models (such as the geocentric model) may continue to be used within their limited domains, even after better models are available. Albert Einstein put it this way: "Every theory is killed sooner or later. . . . But if the theory has good in it, that good is embodied and continued in the next theory."[17] We would add that sometimes the intellectually "killed" model sustains a limited prestige as it coexists with the newer ideas.

As a fourth general feature we note that models are valorized by their utility. We can "objectively" test the domain of validity and hence the utility of a model by comparing its predictions with observations of nature. It may be in this sense that scientific models most profoundly differ from "artistic" creations. Both sorts of creations may be arbitrary in the sense that they are not the only possible result of human creative effort; but their valorization is achieved by differing procedures. Scientists will agree universally on this: a model that does not measure up to observations of the world has reached its limits of validity. No such agreement would be possible in the arts because the procedure for making an appropriate (that is, "objective") measurement may not exist or may not be universally accepted. The "objective" experimental techniques of the scientist constitute a methodology for measuring scientific models against observations of nature and for providing their valorization.[18]

[17] Quoted in Ronald W. Clark, *Einstein: The Life and Times* (New York: Avon Books, 1971), p. 757.

[18] We have been careful to surround the word "objective" in quotation marks. The sense in which we are using this word should be clear, but some philosophers of science argue that it is a loaded term whose meaning is open to dispute.

Because scientific models evolve and have a limited domain of validity, our fifth point is that one must be exceedingly careful about using terms such as ''scientific truth''; whether or not science is able to produce truth in any absolute sense is problematic. For example, what is the *true astronomical reality* of the planets' situation in the solar system? Which model do we take as ''real''? The latest and—presumably—the most accurate, or the next model that will have an even wider domain of validity? Indeed, one can even argue about the absolute character of certain sorts of ''hard facts'' or of measurements. One might feel completely comfortable reporting a count of the number of legs on a table as a ''fact'' about that table. But what about the ''exact length'' of the table? Given the most cunning means of length measurement imaginable, physicists now recognize that the very term ''length'' is ill-defined, for the table's extent may ultimately be determined by the atoms that constitute the table's substance. Quantum mechanics, the best available model for describing atoms, shows that the extent of an atom is best described as a fuzzy (that is, ill-defined) ''cloud'' of electrons in space. One is simply unable to determine the location of the table's edge with unlimited accuracy.

Finally, we remark that scientific models are based upon belief in the validity of certain key assumptions. The most fundamental of these is that the world that the model attempts to describe actually does exist, that the scientist is not simply dreaming it all. One cannot prove the validity of this assumption, of course, and so one accepts on faith the ontological basis for scientific work.

Scientists also assume the validity of human reason or logic—that is, the method of reasoning that we apply to phenomena must be assumed valid, but there may be circumstances where the applicability of human reason can be called into question or where the application of our logic leads us into a ''paradox'' (that is, a situation perfectly consistent with nature, but inconsistent with our rational analysis of nature). We will suggest such a paradox in our discussion of consequences of Einstein's 1905 theory of relativity in chapter 4.

2

THE CLASSICAL BACKGROUND

Then ye how now on heavenly nectar fare,
Come celebrate with me in song the name
Of Newton, to the Muses dear; for he
Unlocked the hidden treasuries of Truth:
So richly through his mind had Phoebus cast
The Radiance of his own divinity.
Nearer the gods no mortal may approach.

—The ode dedicated to Newton by
Edmund Halley in *The Principia*

2.1 NEWTONIAN PHYSICS

If Einstein's theory of relativity seems strange to us, it is largely because we remain Newtonians in our conception of the physical world. To discuss relativity theory and its impact on our established world view, we must therefore first consider Newton's contribution to our intellectual heritage, especially Newtonian mechanics or "classical" mechanics, as it is called.[1] Newton set forth his model of mechanics in his *Philosophiae Naturalis Principia Mathematica* (usually referred to as "the *Principia*"). The publication of this book in 1687 was one of the most significant events in the history of Western culture.[2] Alexander Pope reflected the deep feelings of Newton's heirs in his famous couplet:

---

[1] The adjective "classical" is now applied to all physics that preceded the developments of quantum theory and relativity theory. Modern students still begin their training in physical science by learning classical physics; quantum theory and relativity theory are taught as developments from the established classical (or Newtonian) world view.

[2] The standard translation of the *Principia* is Florian Cajori, *Sir Isaac Newton's Mathematical Principles of Natural Philosophy and His System of the World* (Berkeley: University of California Press, 1966). All references to the *Principia* will be to this edition. Newton prepared 3 editions of the book, the last in 1726.

> Nature and Nature's laws lay hid in night:
> God said, *Let Newton be!* and all was light.[3]

In the *Principia* Newton established what was to be the foundation of modern physical science. It was not until the first decade of the twentieth century that Albert Einstein shook that foundation; and despite Einstein's revelation, large portions of contemporary courses in physics are still based directly on ideas developed or first introduced in the *Principia*.

In the following sections of this chapter we will discuss the meaning of certain key ideas in classical mechanics as the *Principia* presents it. Our special concern will be with Newton's introductory discussion of space and time and with the famous model or three ''laws'' of motion, laws that form the basis of our Newtonian world view.

2.2 SPACE, TIME, AND MOTION

Newton carefully distinguished between ideal notions of space and time (which he called ''absolute'' space and time) and the related measurements (which, it is important to observe, he called ''relative'' space and time) that we make in the course of performing experiments. To quote from the *Principia*:

> Absolute space, in its own nature, without relation to anything external, remains always similar and immovable. Relative space is some movable dimension or measure of the absolute spaces; which our senses determine by its position to bodies; and which is commonly taken for immovable space. . . .
> Absolute . . . time, of itself, and from its own nature, flows equably without relation to anything external, and by another name is called duration: relative . . . time, is some sensible and external (whether accurate or unequable) measure of duration by the means of motion, which is commonly used instead of true time.[4]

After making these distinctions between relative and absolute time and space, Newton went on to distinguish absolute and relative motion in the following words:

> Absolute motion is the translation of a body from one absolute place into

[3] Alexander Pope, ''Epitaph Intended for Sir Isaac Newton,'' in *The Complete Poetical Works of Pope* (Cambridge, Mass.: Houghton Mifflin Company, 1931), p. 135.
[4] *Principia*, p. 6.

another; and relative motion, the translation from one relative place into another.[5]

The idea of relative motion is an important one both for Newton and, later, for the theory of relativity. Let us use a slightly modified version of an illustration Newton gave in the *Principia* to provide a clear picture of exactly what is meant by relative motion. We suppose, for reasons to be discussed later, that all of the motions in this illustration are "uniform," meaning that they do not involve a change in speed or direction of travel. By way of introducing a necessary technical term, we note that changes in speed or changes in direction of motion are given the common name "acceleration."

Bearing in mind that we are considering uniform (unaccelerated) motions, imagine the ship sailing on the ocean, as sketched in figure 2.1. There is a cabin at the rear of the ship and a sailor walking the deck. In order to measure the motion of the sailor, we need to define some point of reference from which to measure his position. Any reference point will do as long as it is well defined. Suppose we choose points along the edge of the ship's cabin to define our frame of reference for measuring distances. We can then measure the distance separating the sailor from the edge of the cabin as shown in figure 2.2.

If we have a clock (to measure Newton's version of relative time) we can also note the time at which the sailor has moved to various distances from the edge of the cabin (figure 2.3). In this way we can measure two things about the sailor: his position in relative space at various relative times. We may then divide the distance he has moved from the edge of the cabin by the time required for him to have moved that distance and so calculate the average speed of the sailor:

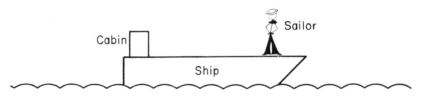

FIGURE 2.1

[5] *Principia*, p. 7; in these statements, the physicists' term "translation" is used to indicate that the motion of an object takes place with no rotation.

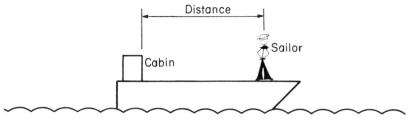

FIGURE 2.2

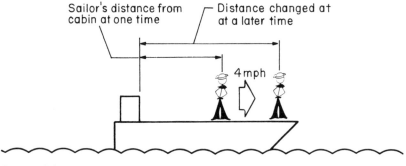

FIGURE 2.3

AVERAGE SPEED OF SAILOR EQUALS
DISTANCE MOVED DIVIDED BY
THE TIME REQUIRED FOR THE MOVE.

The number resulting from this long division measures the sailor's uniform motion with respect to the edge of the ship's cabin; it is a measure of relative motion. Note carefully that the number has no meaning unless a frame of reference is specified along with the number; that is, we need to know the reference frame relative to which the distances were measured in order to understand the meaning of the measured motion. Since we have used the edge of the ship's cabin as our spatial reference frame, we make the average speed number meaningful by saying, ''The sailor's speed is 4 mph *with respect to* (or *relative to*) the edge of the ship's cabin.'' Because the cabin is rigidly fixed to the ship, if the sailor moves at 4 mph with respect to the cabin, he must appear to move at this same speed with respect to any other part of the ship. Therefore, one could also say: ''The sailor's speed is 4 mph with respect to the ship.''

Imagine further that the ship passes a buoy anchored to the ocean floor (figure 2.4). We can speak about the speed of the ship relative to the buoy if we use the buoy as a point of reference from which to measure distances. Suppose the speed of the ship relative to the buoy comes out to be 4 mph due east, as shown in the figure.

Let us now use another frame of reference to measure the speed of the ship: the sailor. At first this may seem to be an unwise choice. After all, the sailor is walking across the ship's deck, and one may rightfully ask whether or not the use of a "moving" reference frame will cause troubles in specifying motion. The answer is No. Our use of the ship as a reference frame seemed appropriate enough, and the ship was found to be moving with respect to a buoy anchored to the bottom of the sea; in fact, as we now know, the sea itself and the entire earth move through space with respect to the sun and stars, and yet we commonly use the body of the earth as a frame of reference. In figure 2.3 the sailor was walking away from the cabin; now suppose that he turns around and walks toward the cabin so that he is walking due west (figure 2.5). If the sailor's motion is due west at 4 mph with respect to the ship, then the ship's speed must be due east at 4 mph with respect to the sailor.

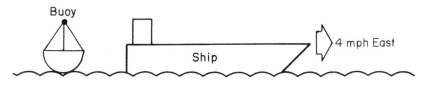

FIGURE 2.4

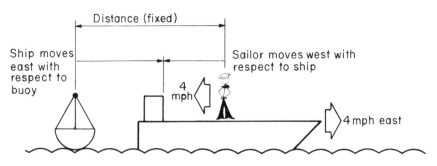

FIGURE 2.5

What about the sailor's speed with respect to the buoy? Recall that the ship has a speed due east at 4 mph with respect to the buoy; therefore, the buoy must have a speed of 4 mph due west with respect to the ship. But this is the same speed the sailor has with respect to the ship; and so the sailor and the buoy seem to be at rest with respect to one another. An observer stationary on the buoy would agree. The sailor would appear to be walking across the deck of the ship, but because the deck is moving under the sailor's feet in the opposite direction, the sailor would not change his distance from the buoy.

The point of this example is to illustrate that in practice we do not really speak of absolute rest or absolute motion or absolute distance. Distance and motion are always measured with respect to some specified or assumed frame of reference. Sometimes the choice of a frame of reference seems obvious, sometimes a number of possible choices occur; but, as Newton points out, we must bear in mind that all are arbitrary choices of convenience. For example, the sailor would probably choose to measure his running with respect to the ship under his feet. On the other hand, the ship's captain would probably choose the speed of the ship with respect to the buoy as the best record for his log. But the ship's captain could have specified the sailor as his reference frame with equal validity, or the sailor could have specified the buoy. The measured speed would be just as meaningful a number in any case.

But what about *absolute* space, motion, and time? Can one identify some frame of reference that may be said to be in a state of absolute rest? Similarly, is there some temporal standard (an absolute clock) that may be applied to measurements of objects throughout the entire universe and so constitute an absolute system of time measurement? A careful reading of Newton's own statements in the *Principia* distinguishing between absolute and relative time and space suggests that he might well have had something very definite in mind when he spoke of absolute space and absolute time.[6] Yet at the same time, he was careful to point out at great length that there is no practical sense in which either of these absolute quantities may be determined. For example, he clearly recognized the possibility

> that there is no such thing as an equable motion, whereby time may be accurately measured. All [real] motions may be accelerated and retarded, but the flowing of absolute time is not liable to any change.[7]

[6] The "Scholium" in the *Principia*, pp. 6–12.
[7] *Principia*, p. 8.

So a truly absolute clock might not actually exist, even though absolute time does.[8]

With regard to motion, Newton pointed out that in practice one can easily determine when two objects (such as the ship and the sailor in our previous example) are in relative motion: if the measured distance separating the two objects is a changing quantity, then the objects are in relative motion. But whether one or the other (or both) of the bodies is in a state of absolute rest or of absolute motion cannot be determined.[9] As Newton's own nautical example illustrates, the only way to determine motion is with respect to some well-defined frame of reference, and there is no way of determining the state of motion of that defined reference frame except with respect to some other chosen reference frame. The process of seeking an absolute reference frame becomes a hopeless chase. "And so," Newton states,

> instead of absolute places and motions, we use relative ones; and that without any inconvenience in common affairs. . . . For it may be that there is no body really at rest [that is, in a state of absolute rest], to which the places and motions of others may be referred.[10]

For Newton, then, absolute space and time (and hence absolute motion) are entities that are not recognizable by any sort of actual measurement or observation. Is there, however, something that can be identified a priori in a state of absolute rest or at some absolute place and at some determined point in absolute time? There can be no question that some of Newton's followers did identify the framework of "absolute space" and "absolute time" with the mind and being of God. In particular, Henry More and Joseph Raphson (contemporaries of Newton and champions of "natural theology," which they sought to substantiate through Newton's models), and Samuel Clark (who many historians believe was serving as Newton's spokesman in this matter) explicitly identify God as comprehending the absolute frames of space and time. There is some debate among scholars, documented in numerous works,[11] about Newton's own views concerning

---

[8] Stephen Toulmin has noted that this distinction between absolute and relative times and positions is analogous to that in geometry between such ideal concepts as "line" and "circle," and practical approximations of these in the fabrications of objects. Stephen Toulmin, "Criticism in the History of Science: Newton on Absolute Space, Time, and Motion, I," *Philosophical Review* 68 (1959): 17.

[9] As we will see, this statement is only valid in situations where the motion is uniform, that is, in unaccelerated motion such as we have stipulated for our present discussion.

[10] *Principia*, p. 8.

[11] See, for example, Alexandre Koyré, *From the Closed World to the Infinite Universe*

absolute space and time. The consensus seems to be that Newton did believe in God as the ultimate, absolute temporal and spatial reference frame.[12]

But whether or not the intellectual underpinnings were based upon religious absolutes, one thing may be said without hesitation about Newton's model of motion: it worked splendidly. And because it was so successful one had to give credence to its underlying axioms. In fact, the model does work whatever one supposes to be the status of absolute space and time, because these notions are simply not needed in its application to the physical world. As we have already pointed out, in a practical sense one can only measure *relative* space and time, and so Newton framed his laws of motion in these terms. However, to the extent that one could ascribe a theological basis to the *Principia*, the success of the model appears to support a theistic world view. Some of Newton's followers used his successful model in just this way. One could even argue that because nature followed the simple and beautiful laws Newton published in the *Principia* (to be discussed in detail in the following section), only the hand of a Divine Architect could have arranged things so. In other words, because the simple laws of Newton worked so well, one had natural evidence for the existence of an intelligible God.

We will, incidentally, find the preceding discussion of relative and absolute motion to be very useful later on when we come to speak of Albert Einstein's modification of Newton's work. As we will see in chapter 3, Einstein turns Newton's boat into a railroad train, and makes excess baggage of the ideas of absolute space and time.

2.3 THE THREE LAWS OF MOTION

Let us now turn to the models, or laws of motion, presented in the *Principia*, laws constituting the framework of Newtonian mechanics that Einstein was forced to supplant. Among specialists there is some controversy

---

(Baltimore: The Johns Hopkins University Press, 1968), chapter 10; and Stephen Toulmin, "Criticism in the History of Science: Newton on Absolute Space, Time, and Motion," pp. 1–29 and 203–227.

[12] Newton's concepts of absolute space and time were attacked in his own day (see Jeremy Bernstein, *Einstein* [New York: Penguin Books, 1973], pp. 129–130) and by Ernst Mach in the nineteenth century. Mach (1836–1916) believed that physics (and philosophy) must be purged of all ideas and concepts not directly observable: all concepts must be defined in terms of direct sense experience. Mach's book, *The Science of Mechanics* (1883) attacked Newton's absolute time and space as undefinable and hence without place in physics. Einstein admired Mach's book, which "exercised a profound influence upon me . . . while I was a

over the logical relationship between the three laws, and in particular whether the first law can be interpreted as a special case of the second.[13] This controversy, however, has little to do with our presentation, whose purpose is to discuss the significance and the utility of the physical science that developed from the laws. While the logical connection between the laws may not be completely clear, the power of their application to physical situations in nature is unquestioned. Here, then, is the statement of the laws as given by Newton in the *Principia*:

LAW 1
Every body continues in its state of rest, or of uniform motion in a right line, unless it is compelled to change that state by forces impressed upon it.

(The use of the word ''right'' in this and the following law stems from the translation of the original Latin text. The modern meaning of the word is ''straight.'')

LAW 2
The change of motion is proportional to the motive force impressed; and is made in the direction of the right line in which that force is impressed.

LAW 3
To every action there is always opposed an equal reaction: or, the mutual actions of two bodies upon each other are always equal, and directed to contrary parts.[14]

(The phrase ''directed to contrary parts'' means ''oppositely directed''; again, the curious phrase is due to the translation of the Latin.)

2.4 THE MEANING OF NEWTON'S FIRST LAW OF MOTION

The first law states that an object initially at rest will continue to be at rest unless acted upon by a force. Furthermore, an object initially in motion

---

student. I see Mach's greatness in his incorruptible skepticism and independence'' (Schilpp, *Albert Einstein: Philosopher-Scientist*, I, 21). For an excellent account of Mach's attack on Newtonian absolutes and his influence on Einstein, see Bernstein, *Einstein*, pp. 130–133.

[13] See, for example, Leonard Eisenbud, ''On the Classical Laws of Motion,'' *American Journal of Physics* 26 (1958): 144–159.

[14] *Principia*, p. 13.

at some speed will continue to move at that same speed and in a straight line (a condition of motion we earlier termed "uniform") unless acted upon by a force. In one sense this law is a statement of cause and effect. The normal state of motion of an object is identified as either a state of no motion at all (and here we note again that all statements about motion must refer to relative motion) or a state of uniform motion (unchanging speed in a straight line). If an object does not follow this normal state of motion, then there must be some cause, and that cause is called a *force*.

To one not trained in the physics of Newton, the validity of this law is not at all obvious; in fact, it is downright contradictory to our everyday experience. Try the following experiment: place any small object on a table top. Observe that the natural state of motion of that object is a state of rest. So far this is consistent with Newton's first law. Now apply a force to the object by pushing it with the tip of your finger and observe that it moves; again, this is consistent with Newton's first law: application of a force has caused the state of motion of the object to change. But now discontinue all contact with the object so that you no longer apply a force to it. According to Newton's law, the object should maintain the speed it had attained at the instant that the force was removed, and it should continue to move at that speed in a straight line. You do, in fact, observe the motion of the object to be (more or less) in a straight line, but the speed is not unchanged. The object soon comes to rest on the table top, something in apparent contradiction to Newton's law.

However, there is no contradiction if we introduce the concept of another force acting on the object, a force called "friction." When you pushed the object across the table top with your finger, you obtained just the result predicted by Newton: the object responded to the applied force and changed its state of motion. When you lost physical contact with the object, you no longer exerted a force. According to your experience, the object should come to rest. According to Newton, if no forces act on the object it should never come to rest. Your experience is reconciled with Newton's first law by the recognition of a force acting on the object, even when your finger no longer touches it. There is, according to the Newtonian scientist, a frictional force between the top of the table and the surface of the object in contact with the table top. That force of friction acts as long as the object is sliding across the table, and because that force acts, it changes the motion of the object in accordance with Newton's law. In fact, a frictional force will always act to slow down the object and, ultimately, to bring it to rest.

Galileo (who arrived at the view expressed by Newton's first law some years before Newton's birth and whose work in mechanics contributed to Newton's development of his model) suggested a way of demonstrating the validity of this analysis. Make the surfaces of the table top and the object more and more smooth by polishing them (or apply a lubricant to the surfaces to fill in all of the microscopic irregularities of the surface and effectively render it more smooth). Try the experiment again and you will see that the object moves for longer and longer periods of time before coming to rest after your finger has stopped pushing. In principle, reasoned Galileo, if the surfaces could be made perfectly smooth and if the table top were long enough, the object would move forever, a conclusion in accordance with Newton's law. So when carefully applied, Newton's first law is found to be consistent with experience in the everyday world. When Newton specifies in the law that *no* force acts on an object, he means not only those forces that we may choose to exert by our own muscle power or by means of some other device; we must be very sure that no forces whatever are at work, including friction and other forces over which we may have little or no control.

We have just discussed Newton's first law of motion as a statement of cause and effect. It is important for our understanding of relativity to be aware that in another sense the law is a specification of the sort of reference frame of measurement in which Newton's laws will be valid, for his laws will not hold in all reference frames. If one's frame of reference is undergoing nonuniform motion (an acceleration), then the motion of objects that is referred to some point in the accelerating reference frame cannot follow the first law (or for that matter the second law, as discussed in the next section).

Suppose, for example, that you are in a car that is speeding up along a straight road. Ahead of you a ball is lying in the road and you measure the ball's state of motion with respect to (or from the frame of reference of) your car. You would find that the ball is moving toward you at an ever increasing speed; that is, the ball appears to be accelerating toward you. Yet there is no force acting on the ball to cause it to speed up, and so Newton's first law is violated. On the other hand, a person standing by the side of the road sees the ball at rest and since no force is acting on the ball, Newton's law is satisfied.

Frames of reference within which Newton's laws are valid are termed "inertial" reference frames (a term introduced in the late nineteenth century);[15] reference frames in which Newton's first law is not valid are called

[15] Toulmin, "Criticism," p. 16.

"noninertial." As we have demonstrated with the ball on the road, inertial reference frames cannot be undergoing an acceleration. But wait. We have just spoken of "an acceleration" without stating its relation to any frame of reference. We have not identified a reference frame with respect to which the acceleration can be said to take place. Nevertheless, our statement about the noninertial character of an accelerated reference frame has validity. We must explain why in detail.

Recall that in our earlier discussion of relative motion we asserted that it is impossible to distinguish a state of absolute motion. There, we specified that the motions taking place were all uniform, that is, at constant speed and in a constant direction. Now consider what happens when the speed of separation of two reference systems changes. To be specific, return to your car with your foot well down on the gas pedal (appropriately called "the accelerator"); again there is a person sitting by the side of the road watching you pass. From measurements you make in your frame of reference, the person seated by the roadside (as well as the roadway and the surrounding scenery) is approaching you at an ever increasing speed; that is, the seated person is accelerating toward you. On the other hand, the seated person determines from her frame of reference that you are approaching her with the same increasing speed. As before with the sailor and the boat, each observer appears to be doing the same thing from the perspective of the other. Before, with uniform motion, we concluded that to decide who was "really" moving (that is, deciding who was in an absolute state of motion) was impossible. The two uniformly moving frames of reference could not be distinguished in any way, except *relative* to some other reference frame. In contrast, we have a way of distinguishing between your reference frame in the car and the person's at the side of the road. By performing experiments in each frame of reference and checking on the validity of Newton's first law in that frame, one can determine whether or not the frame is inertial by definition of the term "inertial frame." In your frame of reference moving with the car, Newton's first law does not apply; that is, yours is a noninertial frame. The seated person by the road, on the other hand, is in an inertial frame because she finds that Newton's first law is valid in her frame. And so there is a qualitative distinction between inertial and noninertial frames. This point bothered Einstein, and we will come back to consider it again in some detail in chapter 5, when we discuss the general theory of relativity.

Let us emphasize the points we have just made. Inertial reference frames are frames in which Newton's first law holds. Any frame moving uniformly with respect to a given inertial frame is also an inertial frame. If a reference frame should accelerate with respect to an inertial frame, Newton's first law will not hold in the accelerated frame; even though observers in each frame see the other frame accelerating, the acceleration is "absolute" in the sense that experiments can be performed to determine which frame is noninertial. Incidentally, these experiments can be performed using measurements made entirely inside the car, without any contact with or observation of the outside. For example, a rider in the car can distinguish his "absolute" acceleration (noninertial character) without making any special sorts of measurements at all: as the car accelerates, riders feel "pushed back" in their seats by that acceleration. A person by the road (or in any inertial frame) feels no such force.

So Newton's first law, sometimes called the "law of inertia," is really a statement of conditions in any inertial reference frame, and it defines the sort of reference frame he used in his subsequent analyses.

2.5 THE MEANING OF NEWTON'S SECOND LAW OF MOTION

Newton's second law can be interpreted as a quantitative extension of the cause-and-effect statement given in the first. It states a proportionality between what Newton terms the "change in quantity of motion" of an object and the magnitude of the force acting upon that object. The modern term for "quantity of motion" is "momentum"[16] and it is defined in the following terms:

MOMENTUM EQUALS THE MASS OF AN OBJECT
MULTIPLIED BY THE SPEED OF THE OBJECT

or, expressing the same definition in the language of algebra,

$$\text{MOMENTUM} = \text{M} \times \text{V},$$
where M = the value of the mass of an object, and
V = the value of an object's speed.

Although it is a detail of interest mainly to individuals who apply Newton's laws to mechanical problems, it should be pointed out that the *direction* of

---

[16] Readers unacquainted with scientific terminology are cautioned that some of the terms used in physical science are identical to words used in colloquial English, but they may have very different meanings in the two applications. In this book we will attempt to provide clear definitions of such words in the scientific sense, and we will be careful to use them only in that sense. The glossary is provided to help with this.

an object's motion is also inherent in a specification of its momentum. Therefore, a change in the momentum of an object may be caused by a change in the object's mass, its speed, or its direction of travel.

We will encounter the term "mass" in a number of contexts in this and later chapters. Each time it will be important that we discuss its specialized meaning in that particular context. Within the context of Newton's second law, mass pertains to the object whose change in motion is being described by the law. It measures the property of the object known as "inertia," which, in turn, is a term describing the reluctance an object has to change its state of motion. Later on we will encounter the term "mass" used to describe something entirely different: the propensity of objects to attract one another through action of the force of gravity. In this "gravitational" context, mass is related to (but not at all the same as) the "weight" of an object. For the present, mass should be thought of as the measure of inertia—how much opposition a body presents to having its state of motion changed. The momentum is defined by the product of this measure of inertia with the speed of the object.

In modern terms, then, the second law may be stated as follows:

> *The change in momentum of an object is directly proportional to the size of the force acting on the object,*

where a change in momentum may be due to a change in the object's mass or to an acceleration of the object or to both an acceleration and a change in mass. It bears repeating at this point that an acceleration includes a change in speed, a change in direction of travel, or a combination of both (figure 2.6).

In many cases, the masses of objects do not change during an experiment, at least to an appreciable extent, and so one can often use an alternate, although more specialized, version of the second law:

> *The force acting on an object is directly proportional to the mass of the object multiplied by the rate at which its speed or direction of motion changes,*

or, alternatively,

> *The force acting on an object is directly proportional to the mass of the object multiplied by the acceleration the body undergoes.*

Or, even more briefly (and more familiarly to former students of high school physics):

NO ACCELERATION

Object at rest

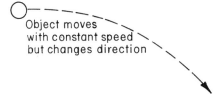

Object moves with constant speed and direction

ACCELERATION

Object moves
with constant speed
but changes direction

Object moves in the same direction but changes
speed

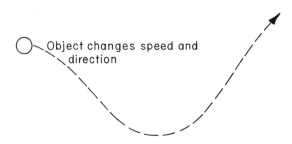

Object changes speed and
direction

FIGURE 2.6

$$F = M \times A,$$

where F = the force applied to the object,
          M = the mass of the object (assumed to be constant), and
          A = the acceleration of the object.

It is important to realize that this law, particularly in its algebraic form, is a quantitative statement of something we know intuitively: the more massive an object (the greater its inertia), the less its response (change in its state of motion as measured by its acceleration) to an applied force.

Consider the following experiment. You place two spheres on a smooth floor. One of the spheres is a bowling ball and the other is a ping-pong ball. You kick the ping-pong ball with a certain force and you see it move from its state of rest (that is, you observe that it accelerates). Now kick the bowling ball with exactly the same force and you will observe that it moves from its initial state of rest considerably less rapidly than did the ping-pong ball. Given your experience at exerting forces on various objects in the world, this result is exactly what you have come to expect. Newton's law quantifies this experience. For a given force (symbolized by F) the product of an object's mass and its acceleration always gives the same number (or, in algebraic terminology, we can say that this product is a "constant"):

$$F = M \times A = CONSTANT,$$

and so when we deal with a ping-pong ball, M is relatively small and A must be large. Conversely, when M is large (as with the bowling ball) A must be correspondingly small so that the product of the large M times A equals the same value of the force.

At this point we can understand why some students of Newton's physics have wondered whether or not the first law is needed at all. Suppose that the object we are considering is at rest or moving uniformly (that is, with zero acceleration). Then Newton's second law states that the force acting upon such an object must be zero; according to the second law, the only way for an object to change its state of motion (that is, for its acceleration to be non-zero) is for a force to act. But these results, obtained from the second law alone, are equivalent to a statement of the first law: the first law appears to be contained within the second.

Much of the power and influence of Newtonian mechanics reside in the second law. It provides a recipe for making quantitative predictions about the physical world. If we know the position, speed, and mass of an object at any given time (the so-called "initial conditions" of an object), and if we further know the values of all forces acting on that object, then we can use the second law to predict the consequences of the actions of those forces in the following way. We first use the second law to calculate the acceleration (the change in speed and direction) that the object will undergo as a result of the forces acting. Given that calculated acceleration, we can predict at any subsequent instant what the new values of the object's speed and position will be. The job is most easily done with calculus—a mathematical tool Newton himself invented for the purpose.

Because Newton's laws are expressed in terms of the most general properties of objects (their motions and masses) and not in terms of the exact nature of the object itself (such as the material from which it is made, its color or age) the domain of validity of those laws was, until the advent of relativity theory and quantum theory in the early twentieth century, thought to be universal. Given the initial conditions of any object, one could apply Newton's model to find its subsequent conditions of motion and position at any time in the future, whether the object studied is a planet, a cannon ball, a billiard ball, an arrow, a stone, or even a human body.

Newton's laws can also be used to investigate the motion of any object at any time in the past as well as in the future. One needs only a specification of the initial conditions of speed, position, and force at a stated time, and then the laws can be "run in reverse" to predict what the values of speed, direction of travel, and position had to have been in the past in order for the present set of observed conditions to come about. Therefore, Newton not only provided a scheme for analyzing the motion of any and all material objects at any place in the universe, he also made it possible to predict motions into the indefinite future as well as into the indefinite past. In principle, one could observe the initial conditions of each and every particle of matter in the universe at the present time and use these to calculate the past and future history of the universe. (Of course, in practice, one would need to have enough computational power and knowledge of all of the forces acting on all of the pieces of matter.) A great deal of research in physics since Newton's time has involved a search for detailed understanding of the forces that can act on objects, because once those forces are understood Newton's laws provide a scheme for determining everything else about the mechanical state of an object that one might want to know.

At present, physicists recognize three main categories of forces in nature: electric-magnetic, gravitational, and nuclear.[17] Newton recognized

[17] For accuracy it should be noted that there are two sorts of nuclear force, the so-called "strong" and "weak" forces. Recently, it has become possible to combine the physical de-

the first two of these (indeed in the *Principia* he gave the first quantitative description of gravitational force) and clearly saw that others might well be found in the future. In his *Opticks* he says:

> For it's well known, that Bodies act one upon another by the Attractions of Gravity, Magnetism, and Electricity; and these Instances shew the Tenor and Course of Nature, and make it not improbable but that there may be more attractive Powers than these.[18]

And so, to Newton and his followers, the phenomenon of motion in nature assumes a profound uniformity and simplicity under the second law:

> And thus Nature will be very comfortable to her self and very simple, performing all the great Motions of the heavenly Bodies by the Attraction of Gravity which interceded those Bodies, and almost all the small ones of their Particles by some other attractive and repelling Powers which intercede the Particles.[19]

With the *Principia*, thinkers obtained the materials necessary for the construction of what has been called a ''world machine,'' a mechanical vision of reality based on the actions of forces upon objects. The consequences of those actions became predictable through Newton's laws. If one could only understand the forces involved in a given situation, one could understand the phenomenon itself and predict its past and future course. That Newton successfully applied his laws not only to terrestrial objects but to celestial phenomena as well showed that his analysis held beyond the earth; indeed, there was no reason to suppose that it would not hold throughout the whole universe. An apparently universal analytical power was within the grasp of the human mind. The success of Newton's approach in physics proved to be an irresistible example for specialists in other fields, so that Newtonian terms begin to appear in a variety of surprising contexts (hence such phrases as ''political pressure,'' ''mental inertia,'' ''economic forces,'' or even Karl Marx's use of economic ''laws of

---

scription of the ''weak'' nuclear force with the electric-magnetic one. Einstein had hoped to combine the descriptions of all of the forces of nature in one ''unified field theory.'' This hope has not yet been fulfilled, although progress at unifying the description of all the forces is being made.

[18] Isaac Newton, *Opticks, or, a Treatise of the Reflections, Refractions, Inflections & Colours of Light* (New York: Dover Publications, Inc., 1952), p. 376. Newton wrote this book in English and the edition cited here includes a foreward by Albert Einstein. The first edition appeared in 1703; the fourth and final appeared in 1730.

[19] Newton, *Opticks*, p. 397.

motion''),[20] although usually without the corresponding laws to make them useful in more than a metaphorical sense.[21]

As an illustration of the sort of analysis that one can carry out using Newton's model, consider the celebrated problem of accounting for the motion of a planet in its orbit about the sun. As we remarked in the previous chapter, the cause of planetary motion had been a persistent mystery from antiquity down to Newton's era. Newton succeeded in applying his model of motion to explain the mystery (although, as we remarked in section 1.7, many years later the limited domain of validity of Newton's model became apparent in this connection). We will assume in our discussion that the planets follow perfectly circular orbits around the sun as shown in figure 2.7. Strictly speaking, planets follow slightly elliptical orbits, but the departure from circularity is very small and, in any case, the argument remains much the same even when the exact shape of the orbits is recognized.

As a planet moves about the sun in its circular path, its orbital speed remains the same but its *direction* of motion is constantly changing (figure 2.8). Because at no point in its orbit does the planet follow even a short segment of a straight line path, we know from Newton's first law that a force must be acting on the planet to deviate it continually from the straight path it would follow in the absence of a force. Given the mass of the planet and the amount by which its direction of motion changes (its acceleration), Newton's second law can be used to calculate both the direction and the magnitude of the applied force. It turns out that the force acting on the planet must always be directed toward the sun (see figure 2.9). Furthermore, the magnitude of the force required to deviate the planet from a straight-line path is exactly that given by the formula Newton obtained for the force of gravity acting between the sun and the planet. Because this force does not act along the direction of the planet's motion, but instead is perpendicular to its motion, the force does not speed up or slow down the planet; it acts only to change the direction of motion.

---

[20] See the discussion in Robert L. Heilbroner, *The Worldly Philosophers* (New York: Simon and Schuster, 1972), p. 161.

[21] There is again some controversy among historians as to whether or not the notion of a world machine, which reached full development in the eighteenth century, was due largely to Newton's work or, perhaps, to the work of others, for example Descartes. See Charles Coulston Gillispie, *The Edge of Objectivity: An Essay in the History of Scientific Ideas* (Princeton, N.J.: Princeton University Press, 1960), p. 92, and Edmund Whittaker, *A History of the Theories of Aether and Electricity* (New York: Harper Torchbooks, 1960), I, 6–7. The fact remains that Newton created the laws describing exactly how the machine must function, even if the idea of the machine itself was established prior to the *Principia*.

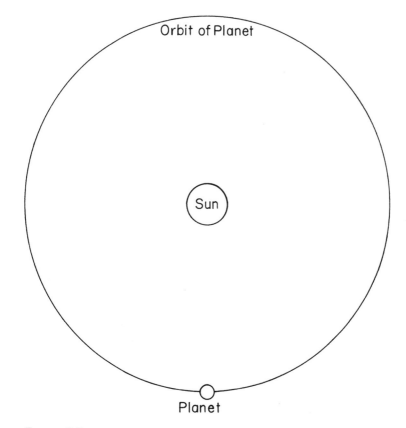

FIGURE 2.7

This analysis provided an explanation for planetary orbits and, in so doing, solved one of the great problems of natural philosophy: what keeps the planets moving in space? Newton's answer is certainly one of the simplest ever proposed, for Newton concludes that *nothing* keeps the planets moving. They tend to keep moving forever, according to his first law, unless acted upon by a force. The force of gravity acts on the planets, but not to change their speed, only to change the direction of their motion and so to keep them moving about the central sun.[22]

---

[22] To be completely accurate, this statement is valid only for the circular orbits we assumed in our discussion. In the actual elliptical orbits, gravity at times does act to speed up the planets and at other times to slow them down. The speeding up at one time is exactly compensated by the slowing down at another time, with the net result that the planet maintains its orbital motion.

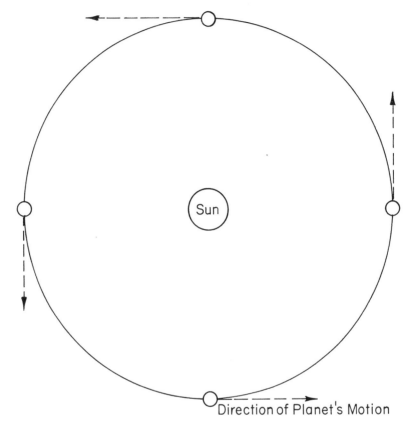

Direction of Planet's Motion

FIGURE 2.8

Actually, Newton's analysis went much farther than this. He was able to combine his laws of motion with his formula for the force of gravity to predict, in complete generality, what should be the motions of planets about a mass like the sun. The result of combining his laws of motion and of gravity was a set of conclusions that were equivalent to Kepler's laws of planetary motion, laws that had been determined from observation prior to Newton's work.

There is an interesting sequel to this story of celestial success. Many years after Newton's death, his models were applied in an attempt to explain slight irregularities in the orbit observed for the planet Uranus. The resulting analysis led to the prediction that a new planet must exist beyond the orbit of Uranus. Given the measured irregularities, Newton's models

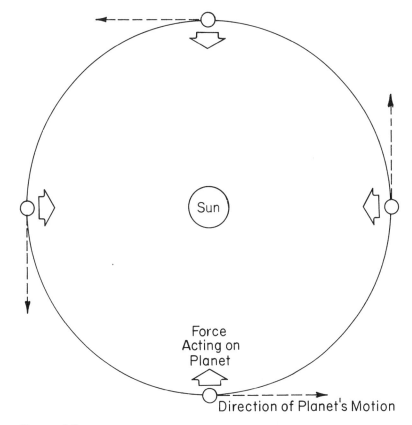

Force
Acting on
Planet

Direction of Planet's Motion

FIGURE 2.9

could predict the position of the new planet in the sky. The new planet, Neptune, was discovered in 1846 by astronomers at the Paris Observatory within hours of receiving the predicted position, providing yet another confirmation of the power of Newtonian physics.

2.6 The Meaning of Newton's Third Law of Motion

Newton's third law of motion is in some respects the most difficult of the three. It asserts that when a force is exerted on an object by some agent, the object exerts an equal and opposing force on the agent. An excellent illustration of the law is provided by Newton himself in the *Principia*: "If you press a stone with your finger, the finger is also pressed by the stone."[23]

If the meaning of this law is not clear, try the experiment suggested here by Newton. Place a stone on a table top and press it with your finger. You will know that you have applied a force to the stone because you can see the stone accelerate as you push it across the table. According to the third law, the stone should apply an equal and opposite force to your finger since your finger is the agent applying the force to the stone. This too you can determine because the nerves in your finger "feel" the force that the stone exerts on your finger.

The third law also permits us to understand the "fictional" force commonly called "centrifugal force." In our Newtonian explanation of planetary motion, some readers may have been surprised to notice that we did not use this term at all. Let us see why. Suppose you have a small ball attached to the end of a string, and you are twirling the ball around in a circular path over your head (we can use figure 2.9 to illustrate this simple terrestrial experiment). If the following discussion of the experiment does not seem correct, tie something to a short piece of string and try twirling it around above your head as you pay close attention to the forces that your hand feels as it holds onto the free end of the string.

In figure 2.9 your hand holding onto the end of the string replaces the sun and the ball attached to the string replaces the planet. As you carry out the experiment you "feel" the ball tugging on the string as though the ball "wants" to pull the string out of your hand. As long as you hold on tightly the ball "stretches out the string" above your head. The force that the ball seems to exert on the string to pull it taut is called a *centrifugal* force (meaning a force that acts to pull something away from the center). This centrifugal force is very real to you as you twirl the ball, and one can explain the phenomenon of the taut string by saying that the ball exerts a centrifugal force on the string to pull it straight.

Similarly, one can analyze the motion of a planet going around the sun by saying that there is a centrifugal force that the planet exerts trying to pull itself away from the sun; at the same time the gravitational attraction of the sun for the planet pulls the planet in the opposite direction, toward the sun (this force exerted by the sun on the planet is called a *centripetal* force—meaning a center-seeking force). In a stable orbit, the centrifugal force is

[23] *Principia*, pp. 13–14.

just balanced by the gravitational (centripetal) force, and so the two oppositely directed forces are in perfect balance; the planet orbits forever in this perpetual state of force equilibrium.

The explanations in the previous two paragraphs sound like reasonable analyses of the terrestrial and celestial versions of figure 2.9. But still we say that in the Newtonian view centrifugal force is fictitious. For one thing, where does this centrifugal force arise? Why does a ball on a string "decide" that it is going to try to pull away from your hand holding onto the other end of the string? Why does the planet "decide" that it is going to pull away from the sun? Of course the planet and the ball do not "decide" to exert forces. The fact is that the centrifugal force you feel acting on your hand is just the equal and opposite force (required by Newton's third law) that the ball exerts on your hand because you decide to exert a force on the ball. Again refer to figure 2.9. The ball on the string is not moving in a straight line, and so it is being accelerated. We know by Newton's second law that this acceleration must be due to a force acting on the ball, and a detailed mathematical analysis shows that this force must act toward the center of the ball's circular path, that is, along the string tied to the ball. This force can have only one origin: the muscles in your hand. You decide to pull on the string; the string, in turn, carries that force to the ball. As a result of the force acting on the ball, the ball deviates from uniform motion and follows a circular path, just as the planet does when acted upon by the sun's gravity. But by Newton's third law, if you exert a force on the ball, the ball must exert an equal and opposite force on your hand, and it does. You feel this force as the "tug" of the ball on the string you hold. It is not necessary to posit a centrifugal force. What is sometimes called a "centrifugal force" is a reflection of the force you are exerting on the ball to keep it in a circular path. Similarly, the sun will feel such a reactive, centrifugal force from each of the planets that it holds in an orbit by its force of gravity.

When we arrive at our discussion of general relativity in chapter 5, we will see that Einstein considerably simplifies the whole analysis of planetary motion; he eliminates the concept of a gravitational force altogether. For the present, we need not discuss Newton's third law any further. In practice, the law is necessary in order to carry out detailed calculations of the effects that forces have upon a system of objects and to obtain other "laws" of mechanics that physicists can use. For our purposes, however, the essence of the predictive power of Newton's model is to be found in the second law.

2.7 MAXWELL'S MODEL OF ELECTROMAGNETIC PHENOMENA

We have devoted a considerable portion of this chapter to a discussion of Newton's models of mechanics because they represent the foundations of physics up to Einstein's day. Before we move on to describe what Einstein did to supplant Newton's work, we must acknowledge some of the contributions to classical physics of the Scottish physicist James Clerk Maxwell (1831–1879), for if Newton provided the background for Einstein's work, Maxwell provided Einstein with the impetus.

Maxwell pondered the results of centuries of experimentation with electric and magnetic phenomena (the experimental work of the brilliant physicist Michael Faraday [1791–1867] was especially important). He managed to synthesize those results into a comprehensive model of electricity and magnetism, or "electromagnetism" as it is now called. The model which Maxwell published in the 1860s is embodied in four equations. These need not concern us directly here but are given a thorough explanation in appendix D.[24] Maxwell's equations rank with Newton's laws among the most honored statements in physics. They serve to describe all of the electric and magnetic phenomena commonly encountered and to reveal the fundamental interrelatedness of electric and magnetic forces. For example, they explain experiments which show how electric phenomena can produce magnetic phenomena and vice versa.

In addition, Maxwell combined his equations to show that this interrelatedness of electric and magnetic phenomena could create a "perpetual motion" of electricity and magnetism in space in the form of waves. In general terms the idea is this. Electric variations produce magnetic variations; but the magnetic variations can produce electric variations, which in turn produce magnetic variations, and so forth in an electromagnetic perpetual motion. As we have already said, these variations move through space as waves, called "electromagnetic waves." Furthermore, from his equations Maxwell was able to predict the speed of those waves. That speed turns out to be the measured speed of light.

Here was an obvious suggestion that light is an example of an electromagnetic wave, a wave consisting of interrelated electric and magnetic variations. Subsequent work has shown that this is indeed the case, and so Maxwell's equations became the basis for the study of optics as well as of electricity and magnetism. And light is not the only manifestation of elec-

---

[24] We recommend reading chapter 3 before reading appendix D.

tromagnetic waves. If the waves have a very low rate of vibration (or, technically, if they have a low frequency or "pitch"), they are radio waves of exactly the sort used today in communication; at slightly higher vibration rates the waves manifest themselves, in turn, as infra-red waves, visible light, ultra-violet, x-rays, and, finally, gamma-rays. This wide array of waves is called the electromagnetic spectrum and is illustrated in box 2.1. Again, all of these different waves are really the same thing: electromagnetic waves, differing only in the rate at which the waves vibrate.

What has all of this to do with the prelude to Einstein's work on relativity? For one thing, Maxwell's equations indicate that according to an observer in a vacuum, electromagnetic waves should only move at one speed, no matter how fast the source of those waves might move with respect to the observer (this prediction of Maxwell's model is discussed in greater detail in appendix D.4). We won't make much of this fact now, but we mention it because it was of crucial importance to Einstein, as we will explain in the following chapter.

Second, the existence of these electromagnetic waves raised the fundamental question of a medium to carry them. We may have discovered electromagnetic waves, but electromagnetic waves in *what*? Waves were a well-known phenomenon in physics, thoroughly analyzed using Newton's

Box 2.1
THE ELECTROMAGNETIC SPECTRUM

Manifestation of the Electromagnetic Wave	*Rate of Vibration of the Wave*
Radio waves	Low
Infra-red waves	Vibration
Red light	Rate
Yellow light	
Green light	↓
Blue light	
Violet light	
Ultra-violet waves	High
X-rays	Vibration
Gamma rays	Rate

models. There were waves on water, waves on violin strings, waves on the taut surfaces of kettle drums; but these waves—indeed, presumably any wave—must be a wave on or in something. Some medium was needed to carry the wave motion. The question after Maxwell's work was, What is the medium that carries electromagnetic waves?

This hypothetical medium was given the name "ether" (or sometimes "luminiferous ether"), and as the validity and power of Maxwell's analysis became evident, "ether physics" became an important line of research even for Maxwell himself.[25] Here are some of the conclusions and attendant problems of ether physics. First, since electromagnetic waves travel through the vacuum of interstellar space as starlight, the ether must pervade all of astronomical space that we can see. But planets, satellites, and stars are not retarded in their motion by any sort of resistance caused by this ether, so it must permit matter to pass right through it with no "frictional drag" of any kind. On the other hand, Newton's physics permits us to infer some of the properties of ether from the form and the speed that we observe for light waves. The ether must, for example, be extremely rigid to be consistent with the observed character of light waves. This high rigidity is to be combined with the property of being so tenuous as to offer no resistance to moving objects.

And there was the question of this omnipresent ether's state of motion. There were suggestions that it could, in fact, provide a physical manifestation of Newton's absolute rest frame. A number of clever schemes were devised to measure earth's speed about the sun in space with respect to the ether; the results of these experiments were startling to physicists: all failed to detect *any* motion with respect to the ether.[26]

Finally, in preparation for our discussion of special relativity in the next chapter, we must take note of the contributions made to this area of physics by the Dutch physicist Hendrik A. Lorentz (1853–1928) because they are of particular relevance to Einstein's work. In the course of attempting to explain the failure to detect any motion of the earth with respect to ether, Lorentz devised a model that involved a set of equations now called the Lo-

[25] A novel about a classical physicist involved in ether research who was unable to adapt to post-Newtonian physics is Russell McCormmach's *Night Thoughts of a Classical Physicist* (New York: Avon Books, 1982).

[26] Probably the most famous example of these experiments was that carried out in the 1880s by Michelson and Morley. A brief explanation of this experiment for the general reader may be found in Banesh Hoffmann, *Relativity and Its Roots* (New York: Scientific American Books, 1983), pp. 75–80.

rentz transformations.[27] He regarded these as formal, mathematical arti-
fices for computation. As we will see in the next chapter, Einstein's theory
of relativity produced a set of equations identical to the Lorentz transformations. These equations comprise the quantitative (as opposed to the conceptual) essence of relativity theory—they are a mathematical statement of
its conclusions. Because Lorentz began his work and published his transformations some years before Einstein's first publication on relativity theory in 1905, the equations retain the name of Lorentz and not of Einstein.
It is possible that Lorentz's results would have been known to Einstein and
exerted some influence on his thinking, but that Einstein did not make specific use of them at that time. Despite the fact that Einstein's theory reproduced the Lorentz transformations as its quantitative "end product," Lorentz's work itself may have had no direct influence on Einstein's development of relativity theory.

Having said this, we hasten to point out that while Lorentz wrote down
the same equations as did Einstein later on, it is not fair to say that Lorentz
in any sense developed or even anticipated relativity theory. He did not.[28]
The path that led Lorentz to the transformations is a complex one, and not
easy to discuss without going into details of physics that are not appropriate
for our treatment.[29] Suffice it to say that Lorentz valued the transformations
as formal, mathematical tools. As we will see in the next chapter, Einstein
arrived at the same set of equations by a clear and logical analysis of fundamental physical phenomena: length and time measurements. When he
had written down the Lorentz transformations, it was not as a computational aid but with full recognition of their physical significance. In short,
Lorentz arrived at the "right" equations but did not recognize their immense physical significance once he obtained them.[30] In any event, Lorentz's model did not succeed in explaining the properties of the ether.

Many strange leopards had entered the temple of science by the begin-

[27] This episode is discussed for the general reader in Hoffmann, *Relativity and Its Roots*,
pp. 81–88.

[28] But he did publish a fine, brief account of Einstein's work in 1919, *The Einstein Theory
of Relativity: A Concise Statement* (New York: Bretano's, 1920).

[29] There is an excellent, although rather technical, discussion of these matters in Abraham
Pais, *Subtle Is the Lord: The Science and the Life of Albert Einstein* (New York: Oxford University Press, 1982), pp. 119–134.

[30] In contrast, the French mathematician and physicist Henri Poincaré seems to have come
quite close to Einstein's analysis some years before Einstein published his relativity theory.
The books by Bernstein, *Einstein*, and Pais, *Subtle Is The Lord*, provide nontechnical and
technical discussions of this point, respectively.

ning of the twentieth century and, in particular, it was in the midst of the rather confused state of affairs regarding the properties of light and ether that Albert Einstein began his work on relativity. We have now sketched enough of the background to begin our discussion of what Einstein did. Although he resolved the problem of the luminiferous ether, we will see that he did not make an attempt to explain its properties. Einstein's answer to the ethereal conundrum is reminiscent of Newton's solution to the problem of the continuing motion of the planets. Newton said *nothing* keeps the planets moving; Einstein said the ether *need not exist*.

3

THE SPECIAL THEORY OF RELATIVITY

God is subtle but he is not malicious.

—A. Einstein

3.1 INTRODUCTION

Albert Einstein's original paper on the theory of relativity appeared in 1905 in an issue of the German journal *Annalen der Physik*. Its title in English, ''On the Electrodynamics of Moving Bodies,'' seems strangely subdued to us now, perhaps identifying a topic without much significance except to a physical scientist.[1] Einstein's scientific output before 1905 amounted to five papers dealing with topics unrelated to relativity. In 1905 he was employed as a ''technical expert third class'' at the Swiss Patent Office in Bern, well away from the mainstream of scientific activities.[2] In other words, the paper burst upon the community of physicists from a most unexpected source. And indeed the exposition of relativity theory in this paper is completely self-contained; it makes no references to other scientific papers, but constitutes by itself one of the clearest presentations of relativity to be found. Its first six pages contain the central ideas of the theory, and we can do no better than to use Einstein's own presentation as a guide

---

[1] The English translation of this paper that will be used throughout this book may be found in H. A. Lorentz, A. Einstein, H. Minkowski, and H. Weyl, *The Principle of Relativity: A Collection of Original Memoirs on the Special and General Theory of Relativity* (New York: Dover Publications, Inc., 1952), pp. 36–65; this text will hereinafter be referred to as *The Principle of Relativity*. The title of Einstein's paper is given a full explanation in our appendix D.

[2] There are numerous biographies of Einstein that deal with this period in his life. We can suggest the brief but very readable one by Bernstein, *Einstein*; for a more thorough biography see Clark, *Einstein: The Life and Times*; for a scientific biography (much of it requiring considerable mathematical sophistication) see Pais, *Subtle Is the Lord*.

for at least the start of our discussion, although later on we will depart from
the order in which Einstein presented certain topics.

3.2 EINSTEIN'S SEARCH: OPERATIONAL DEFINITIONS

Albert Einstein's work on relativity was guided by a series of simple,
even childlike questions. He pursued the answers to these questions re-
morselessly, and in answering them he began to demolish the Newtonian
foundations of physical science. His questions were fundamental: they
were questions of definition—indeed, definition of some of the basic terms
used by scientists and others to describe the world, terms such as "space"
or "time."

Einstein sought definitions that contained implicit instructions for making physical measurements in order to determine the meaning of the terms;
that is, his definitions were so-called "operational definitions,"[3] statements providing guidelines for setting up and carrying out protocols of data
collection. Curiously, his definitions were not stated in mathematical symbols or even in mathematical terms, but as verbalizations. In fact, the first
six pages of Einstein's relativity paper boast no mathematics more complex
than the simple statement that

> *Any object's speed equals the distance traveled by the object di-*
> *vided by the time required to travel that distance,*

or, in algebraic terms,

$$\text{SPEED} \,=\, \text{DISTANCE} \,/\, \text{TIME},$$

an equation that is really a definition of the concept of speed.

Such operational definitions, because they lead to recipes for making
measurements, are at the foundation of modern physical science. They help
give to science what to some is the illusion of an infallible analytic tool, and
such definitions do, in fact, provide much of the power of physical science.

---

[3] One of the leading exponents of the "operational" approach to science, P. W. Bridg-
man, has discussed Einstein's operational definitions in his 1905 theory and contrasts this
approach with that taken later by Einstein in his general relativity theory of 1915, which we
will discuss in chapter 5 ("Einstein's Theories and the Operational Point of View," in
Schilpp, *Albert Einstein: Philosopher-Scientist*, I, 335–354). Incidentally, the operational
approach has its critics. See, for example, Frederick Suppe, *The Structure of Scientific Theories* (Urbana: The University of Illinois Press, 1977), pp. 18–21. The discussions in
Suppe's book are technical, although the pages here mentioned are accessible to general
readers.

In other fields, such as philosophy or literature, verbal definitions are provided by consensus, as they usually are in science, but nonscientific definitions often lack the precision that can arise from an operational definition. For example, one can define such terms as "love" or "God," but disputes arise over the interpretation of the definition, even if a given verbal definition is granted a consensus (which may be impossible for the two terms just cited). In disagreeing about an object's "speed," however, an experiment may always be created to arbitrate. That is, physical scientists agree by consensus that the word "speed" means the distance traveled by an object divided by the time required to travel that distance; they may, therefore, in any situation, use rulers and clocks to measure numbers for the distance traveled and for the time required to travel that distance, and then calculate a speed according to the formula

$$\text{SPEED} = \text{DISTANCE} / \text{TIME}.$$

We say that this sort of definition can create the illusion of infallible analytic power in science because, although it gives terms like "speed" as much quantitative precision as desired and renders disputes resolvable by experiment, in the "real world" such definitions may also be limiting. They may limit precisely because they rule out ambiguity. They also tend to restrict one's vision by suggesting (if not dictating) what aspects of the world we should look at and how we should look, a sort of "preanalysis" imposed by the definitions we use. Furthermore, this sort of definition may not be applicable in many situations. For example, there is no widely accepted operational definition for "love"; it cannot be quantified or tested in a laboratory situation despite attempts by some psychologists to construct mathematical descriptions of behavior.

3.3 THE PROBLEM OF DEFINING TIME

In the definition of speed given above, notice that two other terms were introduced: "time" and "distance," terms whose meaning Newton and his followers had pondered and whose definitions some had ultimately based in a providential world view: God as the ultimate, absolute frame of spatial and temporal reference. Einstein, on the other hand, questioned the meaning of these terms in an operational sense, a sense not specifically distinguished by the providential view. In his 1905 paper he asked what we mean by "time."

To a scientist, and to most nonscientists as well, the most common use of the term "time" is in reference to a number that locates some event or happening. For example, we can say that a plane arrives at an airport at 3:34 P.M. or that two objects collide at 8:27 A.M. These time numbers are thought of as locating happenings or events in a precise way. Einstein recognized this longstanding use but went on to ask how we *define* time in an operational sense: how do we *measure* time?

As a practical matter, of course, Einstein recognized that we measure time numbers by means of devices called clocks, and he understood the term "clock" to mean anything that displays a regularly repeating phenomenon: a heartbeat, the rising and setting of the sun or moon, the vibration of atoms in a substance. The construction or accuracy of clocks was not Einstein's concern, but he was able to see a crucial point about the way we *employ* clocks, and in so doing he arrived at a fundamental insight into the nature of time as we use and measure it. Here is a passage from his original paper on relativity theory:

> We have to take into account that all our judgments in which time plays a part are always judgments of *simultaneous events*. If, for instance, I say, "That train arrives here at 7 o'clock," I mean something like this: "The pointing of the small hand of my watch to 7 and the arrival of the train are simultaneous events.[4]

Although this may sound simple or even trivial, it is a profound observation about the notion of time that we all use. Einstein saw that our measurements of time and our use of these measurements in describing the physical world depend fundamentally upon what we mean by simultaneity.

So pursuing his quest for an operational definition of time, Einstein next asked: What do we mean when we say that two events are simultaneous? In approaching an operational definition of simultaneity, Einstein distinguished two very different situations involving the measurement of time. First, when the clock we are using and the event we are studying are at the same location in space, there is no difficulty understanding what we mean by simultaneity, and hence what we mean by *the* time of an event. When an event occurs, we note the reading of the clock, and that reading is the time of the event. The two entities, the event itself and the clock reading we record to measure the event, are clearly simultaneous. That clock reading, made with a clock at the same location as the event, represents an unambiguous specification of the time of the event.

[4] *The Principle of Relativity*, p. 39.

But Einstein went on to note a more general and not-so-simple case. For events that take place at some distance from the clock we use, or for events that take place while the event or clock is in motion so that the distance between the two is changing, we have a problem. To illustrate the nature of this problem, consider an imaginary experimental situation used by Einstein himself: an experiment carried out on a railroad track of the sort shown in figure 3.1.[5] Suppose that we have a flashbulb at rest on the track. The flashbulb fires and we ask: ''At what time does the bulb go off?'' That is, we seek a specification for *the* time of the event.

If the observer and the clock are right at the location of the flashbulb (sitting on top of it) there is no problem in answering this question, for there is no ambiguity in the time measurement. When the bulb goes off, the observer sitting on top of the bulb records the simultaneous reading of the hands of the clock and that is that.

But if the observer picks up the clock and walks some distance away from the bulb, like observer A in figure 3.1, it takes the light from the bulb a certain amount of time to travel to the observer, and so observer A will give a different time for the flash of the bulb than will an observer on top of the bulb. Observer B still farther away will record the flash of the bulb at an even later time. Here, then, we have three different observers each reporting a different time for the same event. The notion of a single, unambiguous time for the event is challenged.

The point is that once we consider observers separated from an event in space, questions dealing with the time of the event become ambiguous.

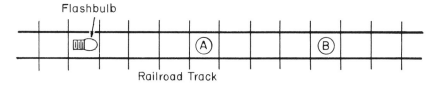

FIGURE 3.1

[5] Einstein did not use a railroad example in his original paper of 1905, but it can be found in his later book on relativity for general readers, Albert Einstein, *Relativity: the Special and the General Theory* (New York: Crown Publishers, Inc., 1961), pp. 21ff. In this book Einstein used strokes of lightning to mark events where we have used flashbulbs as a less providential artifice. We have based our presentation on his later railroad examples instead of the original discussion in the 1905 paper because we find the later work more easily visualized.

Different observers will record different times for the same event. And notice that it is not a question of observers being right or wrong; it is a matter
of difference, and we fully expect different times to be reported. The disagreement among the observers concerning the time of the event has to do
with the finite amount of time required for the information about the event
(the flash) to travel to observers A and B. What about signaling the occurrence of the event in some other way? It turns out that there is no method of
signaling known at present that is faster than light (that is, electromagnetic
waves of the sort predicted by Maxwell). But even if there were such a
super-fast carrier of information, the ambiguity in the measurement of the
time of the event for our three observers would remain. The only way to
remove that ambiguity is to suppose some infinitely fast means of communicating the occurrence of the event. A truly instantaneous means of
communication is unavailable and, as we will discuss in section 4.4, there
are very important reasons for supposing that such instantaneous communication is impossible.

So Einstein concluded from his analysis that when we are dealing with
clocks and events that are separated in space or that are in motion with respect to one another, we must be very careful when we ask for the specification of the time of an event because this question is ambiguous from an
operational point of view.

The discussion in this section of the details of making practical measurements with clocks may appear to be hairsplitting, but it turns out that the
practical questions Einstein raised in this way are crucial. Our traditional
(Newtonian) uses of time and simultaneity suggest a simplicity that
prompted few before Einstein to study carefully their exact operational
meaning. It is a tribute to his genius that he was able to recognize their importance.

3.4 EINSTEIN DEFINES A COMMON TIME SYSTEM

Having recognized the ambiguity inherent in making time measurements
when observers are separated in space from the events they are watching,
Einstein next sought an operational definition for the concept of the time of
an event; that is, he sought some sort of procedure that would permit scientists and others to specify and report a single, unambiguous time for an
event. Here is how Einstein developed this operational definition. Again,
we will use an imaginary situation on a railroad track to illustrate. Consider
in figure 3.2 an observer named Pablo and his clock, both at rest at position

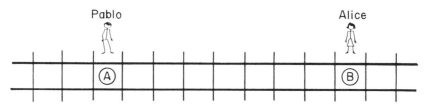

FIGURE 3.2

A on the track. Events at A are easily timed by Pablo: the time of an event at A is defined by Einstein to be the reading of Pablo's clock which is simultaneous with the event at A (in the terminology of physics this time is called the "proper time" of the event).

A separate clock and an observer named Alice are at rest on the track at position B. Alice can use her clock to produce numbers that represent an unambiguous definition of the time of events at B.

So we now have two unambiguous, but independent, definitions of time: Pablo-time and Alice-time, for events taking place at positions A and B, respectively. But is there what Einstein called a *common* system of time that we could define and use to represent the times of events at both A and B and at any other point in space? And if such a common system of time is a possibility, how can observers everywhere synchronize their clocks to read this common time?

Intuitively we suspect that there is such a common time for Pablo and Alice at points A and B and for everyone else. In fact, we think that we use such a common time in our daily lives. In principle, we think that we can synchronize our clocks to read a common system of time like Eastern Standard Time, or Greenwich Mean Time, and so, we trust, can people in Paris, or San Francisco, or Málaga, or anywhere else. So this notion of a common system of time is really quite familiar, and the act of "setting the clock" is really an act of synchronizing a clock to read the same time value as all of the other clocks on the common system. Einstein then asked: How can we actually establish this common time system, given all of the difficulties we encounter when events and clocks are separated in space?

Again we return to the railroad track in figure 3.3 for a thought experiment. The figure now shows Pablo, Alice, and you located on the railroad track. Einstein proposed the following procedure (operational definition) for establishing a common system of time for the three of you and for everyone else at rest along the track. A runner carries instructions to both Pablo and Alice: at exactly noon on Pablo's and Alice's clocks, each is to

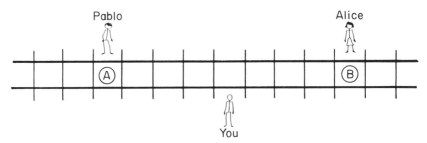

FIGURE 3.3

fire a flashbulb. The runner then carefully measures the distance between Pablo and Alice with a ruler. You stand exactly halfway between the measured positions of Pablo and Alice, and you set up two mirrors (shown in figure 3.4) so that the rays of light from both flashbulbs will be reflected to your eye. In this way you are able to see both Pablo and Alice at the same time without turning your head. Notice, too, that light rays reaching your eye from Pablo and Alice will follow paths that are of exactly equal length.

At noon on Pablo's clock he fires a flashbulb; Alice also fires her flashbulb when her clock reads noon (figure 3.5). If you see both flashes simultaneously in your mirrors, you can conclude—and report to Pablo and Alice—that their clocks are synchronized: they both read the same time.

If the flashes do not reach you at the same time, you can tell the person whose flash reached you first that her or his clock is faster than the other person's and that the rate of that clock should be adjusted; you can then repeat the whole experiment over and over again until both flashes reach your eye at the same time. The two clocks will then be synchronized. In fact,

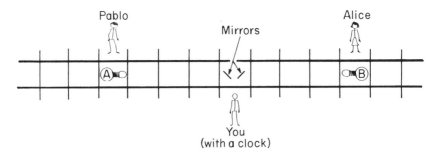

FIGURE 3.4

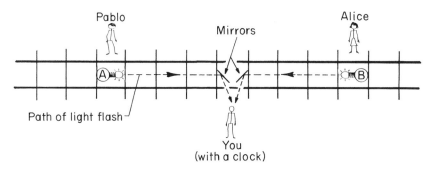

FIGURE 3.5

Pablo and Alice can agree to send out flashes each hour or each minute or each billionth of a second to check continually on their state of synchronization with as much accuracy as required.

In this way it was possible for Einstein to synchronize Pablo-time and Alice-time. Both observers can be assured that their clocks tick at the same rate. We have, in short, an operational definition for the term "synchronized clocks." And what we have done for Pablo and for Alice we can do for as many clocks and observers as we like at rest along the railroad line. As figure 3.6 suggests, in principle we can cover all points along the railroad track with precisely synchronized clocks and so provide an operational definition for a common system of time all along the track.

Having carefully defined the common time system, Einstein was then in a position to define operationally what he meant by the time of an event, even when the event is well separated from an observer. The time of an event was defined by Einstein to be the time given simultaneously with the event by one of the synchronized clocks that happens to be located at exactly the same point in space as the event.

We can use runners or telephones or any other method we choose to report that time for the event to all of the other observers along the track. No-

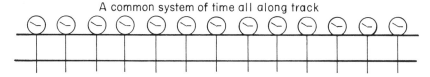

FIGURE 3.6

tice that this unambiguous time for an event will not, in general, be the time at which an observer actually *sees* or otherwise learns of it. The time at which an event is actually seen will depend on the distance of the observer from the event, and so will differ from observer to observer.[6] Nevertheless, Einstein's definition of the time of an event on the common time system is a unique temporal specification.

3.5 THE PROBLEM WITH MOVING OBSERVERS

Given this careful and unambiguous definition of time, Einstein then went on to analyze another simple situation and to ask another very simple question. So far, we have been dealing with observers and events at rest on our railroad track. But now suppose that an observer named Gertrude has a clock of her own and is riding a railroad car at a uniform speed v along the track, as in figure 3.7. Notice we are specifying that Gertrude's speed be uniform, that is, not accelerating. This will be true for all of the moving reference systems discussed in this chapter, and this is the assumption that makes Einstein's relativity theory of 1905 a "special" theory of relativity. His later and "general" theory covers motion that is accelerated, as we will see when we discuss it in chapter 5.

Suppose that Gertrude wants to use her clock in order to measure the times at which various events take place. Because you and your colleagues at rest by the side of the track have already gone to such great effort to establish an extensive network operating on a common system of time, Gertrude decides to synchronize her clock to your common time system so that her time measurements can be compared easily with yours. First, however, she wants to make sure that all of the clocks on your common time system are really synchronized with one another, as you claim. She decides to use the procedure that Einstein specified in his operational definition of synchronization and that you initially used to set up the common time system.

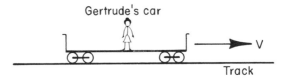

Gertrude's car

Track

FIGURE 3.7

---

[6] This point will be discussed at length in sections 4.7–4.9.

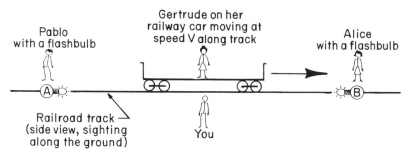

FIGURE 3.8

Again, Pablo and Alice are at rest on the track (figure 3.8). Again they agree to fire flashbulbs at noon according to their clocks on the common system of time. Gertrude is equipped with a mirror apparatus like yours so that she can see signals from both Pablo and Alice without moving her head. The experimental arrangement is exactly as illustrated in figure 3.5, when the common system was first established; now, in addition to you at rest along the track, Gertrude also will be observing the light flashes from her uniformly moving car.

Suppose that at exactly noon on your clock Gertrude passes you on the track. A few instants later the flashes from both Pablo and Alice reach your eye and you see these flashes as simultaneous events, thus confirming that Pablo's and Alice's clocks are still synchronized. But Gertrude disagrees with you. For her the two flashes cannot be simultaneous. Before either flash has reached you she has moved along the track toward Alice and away from Pablo (figure 3.9). Therefore Alice's flash reaches her first. Then, as both flashes reach you, Gertrude has moved even farther along the track,

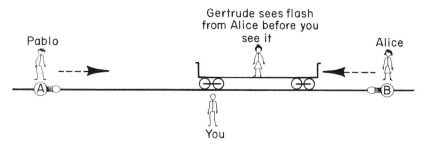

FIGURE 3.9

closer to Alice and farther from Pablo, so that it takes light extra time to reach her from Pablo (figure 3.10).

Thus, events simultaneous to you, at rest on the side of the track, are not simultaneous to Gertrude moving along in her car. The clocks on the common system of time, which appeared to you to be in perfect synchronization, do not appear to be at all synchronous to Gertrude.

Recall that the basis of our ability to synchronize clocks and to establish a common system of time—indeed, the basis of Einstein's very definition of synchronized clocks on the common system of time—is our supposed ability to specify simultaneous events. Since you and Gertrude cannot agree on what is simultaneous, the two of you cannot agree on the synchronization of clocks, and Gertrude will never be able to synchronize her common time system to the common system of time that you use at rest along the side of the track. The same would be true of an observer and clock located anywhere on Gertrude's car and moving along with her. On the other hand, Gertrude could arrange to define *her own* common system of time for use by any and all observers moving with her on the car. All such observers on Gertrude's car would be at rest with respect to one another, just as all of the observers on the ground by the side of the track are at rest with respect to one another, and so are able to synchronize their clocks and to define a common time system. So while you and others at rest along the side of the track, and Gertrude and others riding along on her car can each define perfectly unambiguous and useful common time systems, these two time systems can never be brought into synchronization because observers at rest along the track and observers moving on Gertrude's car cannot agree on simultaneous events. As Einstein put it in his original paper,

We cannot attach any absolute signification to the concept of simultane-

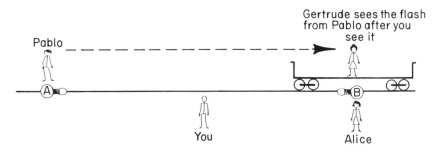

FIGURE 3.10

ity, but that two events which, viewed from a system of co-ordinates, are simultaneous, can no longer be looked upon as simultaneous events when envisaged from a[nother] system which is in motion relatively to that [first] system.[7]

So according to Einstein's series of definitions, we are forced to deny the validity of the notion of an absolute, universal time system. Your common system will differ from the one that Gertrude would set up; her time system cannot be made synchronous with yours because you and she do not see the same events as simultaneous. Notice again that this is not an issue of observers being correct or incorrect in what they see and report. You and Gertrude just differ. There is no privileged or "objective" viewpoint.

At this point in his paper, Einstein goes on to develop his theory in quantitative detail and we will not follow him in that process. We will, however, discuss the ideas underlying that development and, in the following chapter, some consequences of the fully developed theory. First, we must explain the two postulates that lie at the heart of special relativity theory.

3.6 THE TWO POSTULATES OF SPECIAL RELATIVITY THEORY

The first postulate Einstein calls "The Principle of Relativity,"[8] and we state it as follows:

The laws of physics must be the same for all observers moving in nonaccelerated (inertial) reference systems.

This postulate requires some discussion because it has far-reaching consequences both for the 1905 special theory and for the general theory treated in our chapter 5. First, notice that the postulate applies to frames of reference that are not undergoing an acceleration (the "inertial" reference frames mentioned in the last chapter). In chapter 5 we will see how this postulate was extended to cover accelerating systems of reference as well, but for now we limit the discussion to observers at rest or moving uniformly with respect to one another on our railroad track.

At rest by the side of the track, you can formulate models of physics (for

---

[7] *The Principle of Relativity*, pp. 42–43.

[8] Whittaker points out in *A History of the Theories of Aether and Electricity*, II, 30, that the term "principle of relativity" was used in 1904 in an address by the French physicist Henri Poincaré.

example, Newton's laws or Maxwell's equations) and perform experiments to test and extend those models. Gertrude, riding in her car, can do the same thing. The principle of relativity asserts that you must both get the same laws. This does not mean that you would obtain the same numbers measuring a given phenomenon, only that the laws relating those measurements would be the same. This is difficult to state in abstract terms, so let us discuss a specific example.

Suppose that Gertrude is tossing a ball vertically in the air as she rides past you in her railroad car. We will analyze this situation from the standpoint of a Newtonian physicist for purposes of illustrating what we mean by "the Principle of Relativity." (This analysis will be quite accurate if Gertrude's speed is small compared to that of light; that is, for small values of v we can neglect the effects of relativity theory that turn out to vitiate the Newtonian analysis at high speeds.) An observer at rest on the car will see Gertrude standing still and the ball moving up and down along a vertical path, as sketched in figure 3.11. Experiments done on the car reveal that the motion of the ball is described by Newton's second law (discussed in section 2.5). We repeat our statement of this law here, not because we will need to apply it, but because reference to a specific law of physics makes our discussion simpler:

The force applied to the ball is equal to the mass of the ball multiplied by the acceleration of the ball.

Or more abstractly,

$$F = M \times A.$$

Actually, there are two forces at work on the ball, one supplied for a brief time by Gertrude as she lofts the ball, and the steady force of gravity, always acting toward the ground. Newton's law works to describe the effects of both.

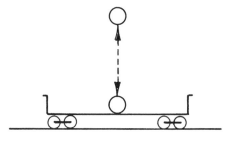

FIGURE 3.11

So far, this is similar to the sort of discussion we had in section 2.4 in connection with Newton's laws of motion. But now look at figure 3.12, which shows the same ball tossing from your vantage point, at rest along the side of the track. As Gertrude moves along with a uniform speed v tossing the ball vertically, you see the ball follow a series of arcs in the air. (Actually, careful measurements will reveal that the ball moves along *parabolic* paths.) These arcs may be understood as the result of combining the vertical motion of the tossed ball with the concurrent horizontal motion of Gertrude and her car to the right at the constant speed v.

You and the rider on Gertrude's car obviously measure different paths for the ball, and the numbers that you report characterizing the ball's motion will differ. You, for one thing, will report a horizontal motion at a steady speed v for the ball, while the rider in the car will report no horizontal motion at all; you will report a position of the ball that changes both horizontally and vertically in time, while Gertrude or her rider will report only a vertical change in position. On the other hand, you will measure an acceleration for the ball (due to gravity) that is identical to the one Gertrude or a rider on her car measures, so that some things that are measured will be common to you and observers on the car. You will also find that Newton's second law describes the motion of the ball, just as it did for riders on Gertrude's car. In other words, you too would write the relation F = M × A to describe the motion.

This is what is meant by the principle of relativity. You and Gertrude may disagree on some numbers that you measure (although you will also agree on others), but the *laws of physics* that you use will be the same.

The principle of relativity also bears on our inability to distinguish states of absolute rest or absolute motion in an inertial reference frame. In section 2.2 we mentioned Newton's conclusion that even if such absolute states existed, in practice we only deal with relative motion. A rider on Gertrude's car can look through a window to see a second car on the adjacent track moving past at some constant speed v. The rider can assert that she and the other train she sees are in relative motion; but can the rider say more? Can

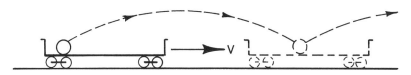

FIGURE 3.12

she say who is "really" moving (that is, can she specify motion in an absolute sense)? The answer, as we observed in chapter 2, is No. The statement of relative motion is all that one can make. By craning her neck at the window and looking down at the track the rider could determine whether her car or the other car or both are in motion relative to the ground, but that would still be a statement of relative motion. Now suppose that all of the window shades in Gertrude's car are pulled down so that riders are unable to observe anything outside the car. All of their measurements and experiments must then relate to phenomena inside the car. The principle of relativity corresponds to the fact that persons riding in Gertrude's car with the window shades pulled down (that is, unable to communicate with the outside) cannot tell whether or not they are moving with respect to any other inertial systems (here, of course, we are assuming that the ride is perfectly uniform and hence smooth) any more than you, as you read these words, sense the fact that you and the earth are hurtling through space at 67,000 mph due to the earth's orbital motion.[9] According to the principle of relativity, because the laws of physics are the same in all inertial systems there is no experiment that can be performed to distinguish those in motion or at rest with respect to any other inertial systems unless relative motion with respect to those other systems is observed directly (by raising a window shade and looking out).

This point is so important to both special and general relativity that it bears further elaboration. Suppose the principle of relativity were incorrect; suppose inertial observers traveling at different speeds actually discovered different laws of physics to describe the same phenomenon. Then as we are about to show, there would be an absolute frame of reference from which to judge motion.[10] Suppose the usual expression for Newton's second law,

$$F = M \times A,$$

were valid only for observers in a particular inertial reference frame. Observers moving uniformly with respect to this frame would determine some different law, some law depending on their relative speed v (for example, Newton's law might become

    [9] To be completely accurate, the earth is not a truly inertial system. By virtue of its orbital motion about the sun it, like the other planets, is undergoing an acceleration. But the amount of the acceleration is so small that for many ordinary purposes the departure of the earth's properties from those of a truly inertial frame of reference is negligible.

    [10] In fact, if the existence of an absolute frame were not implicit in the assertion that the principle of relativity is incorrect, the statement that "the laws of physics depend on the *speed* of inertial observers" would have no meaning.

$$F = M \times A + (\text{SOME NUMBER}) \times V$$

in the moving frame). Even in a car with the shades pulled down over the windows one could tell by experimenting with moving masses when $F = M \times A$ is valid and hence when one is at rest with respect to the "special" inertial frame. This frame then would be "absolute" because it would be the only one in which $F = M \times A$ is valid. All motion, anywhere in the universe, could be referred to the absolute $F = M \times A$-frame of reference. But as we pointed out in section 2.2, Newton realized that no law of physics known to him could distinguish inertial frames of reference—in other words, a law of the sort we have supposed that depends on relative speed v does not exist. This conclusion—the essence of Einstein's first postulate, or the principle of relativity—remains as true as it did in Newton's day.

Banesh Hoffmann paraphrased the content of the postulated principle of relativity very neatly:

> If we are in an unaccelerated vehicle, its motion has no effect on the way things happen inside it.[11]

Identical experiments we perform inside any unaccelerated frames of reference will yield the same results; the detection of absolute (unaccelerated) motion is ruled out. Notice that the statement refers to experiments *inside* the frames (for example, inside the car with the window shades pulled). An experiment involving a view of another reference frame (for example, by looking out the window at the adjacent car passing by) will of course detect *relative* motion (a ball tossed by someone in the next train moves along arcs while a ball you toss moves vertically). The person on the other train, performing the same internal experiments as you, will obtain physical laws identical to yours (she will see her ball move vertically; looking outside at your passing train, she will see your tossed ball move in an arc).

Einstein states the principle of relativity as a postulate, albeit one that seems to be in accord with our experience of uniform motion. But he does not restrict his statement to experiments dealing with motion; he says that *all* laws of physics will be the same for unaccelerated observers. This statement must be a postulate because Einstein cannot prove its validity; he believes it because he thinks that this is the way the world must be, and no law of physics has yet been found to violate his assumption.

It is worth pointing out that Einstein's insistence on the principle of relativity effectively "postulates to death" the popular late nineteenth-cen-

---

[11] Hoffmann, *Relativity and Its Roots*, p. 91.

tury notion of an ether as the medium that carries light waves and with respect to which the speed of light is determined. In section 2.7 we mentioned that Maxwell's equations can be used to calculate the speed of light. If light's speed is defined with respect to an ether, then as an observer moves with respect to this medium at various uniform speeds her measured speed of light must change. But this means that Maxwell's equations, the laws of physics that predict light's speed and do not require any specification of the observer's speed,[12] must also change depending upon the observer's state of motion in the ether, a conclusion in clear violation of the principle of relativity. Thus Einstein says in his original paper,

> The introduction of a "luminiferous ether" will prove to be superfluous inasmuch as the view here to be developed will not require an "absolutely stationary space" provided with special properties.[13]

Einstein solves the "ether problem" by making the whole concept irrelevant. The experiments designed to detect the motion of the earth with respect to the ether that we described in the last chapter were doomed to fail because the ether cannot be detected. Unlike waves of water, sound and waves on taut strings, electromagnetic waves evidently travel without any detectable medium.

Einstein's second postulate may be stated as follows:

> *The speed of light in a vacuum is the same, regardless of the speed of the light source with respect to the observer of the light.*[14]

This too is a postulate. It is something that Einstein assumed to be true; it was to be confirmed or not confirmed, depending upon how well predictions based upon it agreed with what was observed to be the case. In fact, at the time Einstein published his theory (1905) there was reason to believe that this postulate might have experimental and theoretical support. As we observed in the last chapter, near the end of the nineteenth century James

---

[12] That Maxwell's equations do not require the observer's speed to be specified is important here. If the equations involved the speed of the observer, the calculated value for the speed of light might indeed depend on the observer's speed. If the observer's speed does not enter Maxwell's equations at all, then that speed cannot affect the calculated speed for light.

[13] *The Principle of Relativity*, p. 38.

[14] When light moves through material media, such as water, glass, or air, its speed will not be the same as in a vacuum. We should also point out that when Einstein went on to develop his general theory of relativity, he found that this postulate is invalid when the light is viewed from a reference frame that is undergoing an acceleration or when the light travels in a region of space in which gravity is present. Within the assumptions of "special relativity," however, the postulate is confirmed by experiment.

Clerk Maxwell published a set of equations describing the behavior of light and all other forms of electromagnetic radiation. These equations became the basis for all of optical science because they served to predict the behavior of light in all circumstances that could be realized experimentally. Einstein's postulate concerning the constancy of the speed of light is consistent with Maxwell's equations, for Maxwell's equations allow light to travel at only one speed in a vacuum, no matter how the source of the light may move with respect to the observer. (This point is discussed at length along with related aspects of Maxwell's equations in appendix D.) And so the recognized success of Maxwell's equations at describing a wide range of optical phenomena suggested the validity of Einstein's postulate, a postulate that subsequently has been found to be in complete accord with observations.

Here is an example of one such experimental confirmation. If the postulate were incorrect and the speed of light depended upon the speed of the light source, then the effect should be most noticeable for experimental light sources moving past an observer at nearly the speed of light (186,000 miles per second). Such very high speeds can only be achieved experimentally in the particle physics laboratory. In one experiment, reported in 1964, subatomic particles called neutral pi-mesons were found to travel past the observers at more than 99 percent the speed of light. These particles have a short lifetime (about one millionth of a billionth of a second), and when they spontaneously decay they give off a burst of electromagnetic radiation (the radiation is in the form of gamma-rays and not visible light, but it is still electromagnetic radiation). The speed of this radiation was measured, and the value agreed with the speed of light as calculated from Maxwell's equations or as measured when the source of the light is at rest with respect to the observers. As required by Einstein's second postulate, in measuring the speed of light it makes no difference how fast a source of light moves past an observer.[15] But such experimental confirmations are only recent. At the time Einstein asserted this postulate, it was totally unlike anything a Newtonian physicist would have believed.

Indeed it is this postulate that renders the results of relativity theory so difficult to reconcile with our ''common sense'' notions of what the world should be like. When considered carefully, this postulate in itself just does

[15] This work was brought to the authors' attention in A. P. French, *Special Relativity* (New York: W. W. Norton & Company, Inc., 1968), p. 74. This is an excellent introduction to Einstein's 1905 theory for undergraduate physics students. The text abounds with experimental evidence and examples of application of the theory.

not seem to "make sense" to our Newtonian way of looking at the world. It asserts that no matter how the light source moves relative to the observer, the same speed for light will always be measured. This is not the way things usually are. For example, if someone shoots a gun at you, and if you are running away from the gun (or equivalently, if the gun is backing away from you), the faster you and the gun separate, the more slowly the bullet will seem to approach you. In fact, if you and the gun separate from one another at the muzzle velocity, the bullet will appear to you to stand still.

This is the situation for bullets fired from guns. It also works for waves of various sorts, like water waves or sound waves: run away from an approaching wave fast enough, and the wave will seem to stand still. This scenario works in most situations we encounter, and according to Newtonian physics it should always work. But it does not work for light waves. No matter how fast you run away from a source of light, the light will move at the same speed according to your measurements. To a Newtonian scientist, this situation just does not make sense. Nevertheless, this was Einstein's second postulate. Let us now see how he uses it to construct his theory of relativity.

3.7 THE RELATIVITY OF TIME MEASUREMENTS

We will use another thought experiment[16] to see in greater detail just how Gertrude's clock and her common system of time will differ from your clock and your system of time. Suppose that Gertrude has something called a "light clock," sketched in figure 3.13. Two mirrors separated by a distance L are aligned so that they are perfectly parallel to one another. Light detectors are mounted in the surfaces of the mirrors to give an indication each time a pulse of light strikes one of the mirror surfaces. A short pulse of light has been introduced into the clock so that it will bounce back and forth continuously between the mirrors.

Then, if T is the amount of time required for the light to bounce from the bottom mirror to the top mirror and back again in a round trip, we can say that the light covers a distance twice L (or 2L speaking algebraically) in a time T. In general, the speed of light must be given by the formula

---

[16] Again, this is not the one that Einstein used in his original paper on relativity; it has, however, been used by several authors in explanations of relativity theory. For example, see Bernstein, *Einstein*, pp. 63–65, and Hoffmann, *Relativity and Its Roots*, pp. 101–103.

THE SPEED OF THE LIGHT PULSE EQUALS THE DISTANCE COVERED
DIVIDED BY THE TIME FOR THE TRIP,

or symbolically,

SPEED = DISTANCE COVERED / TIME FOR THE TRIP.

If the DISTANCE COVERED = 2L and the TIME FOR THE TRIP = T, this formula may be written more briefly as follows:

SPEED = 2L / T.

Thus we are able to use the light clock to measure the speed of light.

Suppose that Gertrude brings her light clock to the point on the track where you are standing. Since you and she are at rest with respect to one another, you are able to agree on a common system of time. Both of you will measure the same values for L and T, and so you will both get the same value for the speed of light.

Now suppose Gertrude puts the clock in her railway car, positioning it so that the light path is vertical while she moves uniformly past you on the track at a speed v. Einstein's first postulate guarantees that she will be able to use the same laws of physics and hence the same techniques of measurement as you do. What does she measure for the speed of light? Gertrude is at rest with respect to her light clock, and so her measurements should be identical to those made when she was at rest with respect to you at the side of the track (as shown in figure 3.13). As the light pulse makes a round trip,

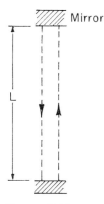

Time for light to make round trip from the top mirror to the bottom mirror and back to the top = T

Speed of light = c

$$c = \frac{\text{Distance travelled in round trip}}{\text{Time for round trip}}$$

$$c = \frac{2L}{T}$$

FIGURE 3.13

Gertrude determines that it covers a distance 2L in time T, and so she again calculates the speed of light to be,

$$\text{SPEED} = \text{DISTANCE COVERED} / \text{TIME FOR THE TRIP}$$
$$= 2L / T,$$

the same result you and Gertrude obtained when the clock was at rest beside you.

What do *you* measure as Gertrude and her clock move past you on the track (figure 3.14)? To you, the light does not move along a path of length L as it moves from the bottom to the top mirror, because while the light moves vertically between the two mirrors, the mirrors themselves move horizontally along the track. To you, therefore, the light bouncing from the bottom mirror to the top mirror will follow a path having a length greater than L.

Notice that the contrast between figures 3.13 and 3.14 is similar to that shown in figures 3.11 and 3.12, where Gertrude was tossing the ball vertically on her moving car. There you saw the ball follow a path that was a (parabolic) arc; here you see the light follow a straight diagonal path. Gertrude of course sees vertical paths in both situations.[17]

Observers stationed at rest along the side of the track with clocks synchronized to their common time system can report to you the time required for the pulse of light in Gertrude's clock to make a round trip. We will symbolize this time as T_G. It is important to emphasize that this number, T_G, is the time for a round trip of the light in Gertrude's clock as measured by observers at rest by the side of the track. This will not be the same as the round-trip time Gertrude measures because she, and her light clock, are moving with respect to the observers at rest along the side of the track.

According to you, as the light moves vertically a distance L, the mirrors also move along the track a distance equal to the speed of the car, v, times half the round-trip time for the light pulse (as measured on your common system). So the total path followed by the light pulse in making a round trip

[17] The difference in path shape for the objects on the car as viewed by someone standing by the side of the track (a parabolic arc for a tossed ball and a straight diagonal line for light) is due to the differing effect of gravity in the two cases. According to classical theory, as well as Einstein's in 1905, the motion of light is not altered by the force of gravity, and so it always moves through space in a straight line, whereas gravity does accelerate the ball and causes it to follow a curved path according to you. We will see in chapter 5 that the conclusion about the straight path of light is not correct, although the error made for light moving near the surface of the earth is nearly negligible.

A Light Clock in Motion

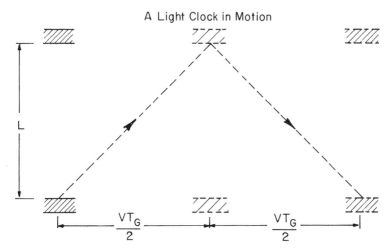

FIGURE 3.14

appears to you to be greater than 2L. In fact, although it is a quantitative detail not central to our discussion, one can calculate the exact length of the path followed by the light from your point of view using the so-called Pythagorean theorem of plane geometry as shown in figure 3.15 (optional).

Box 3.1 (optional) summarizes the calculations of the speed of light that you and Gertrude make. In the Newtonian world, since the light clock ticked once each T seconds when it was at rest with respect to you on the track, you must conclude that the same clock will tick at the very same rate (once each T seconds) as it moves past you on Gertrude's car; therefore, your time measurements and Gertrude's should be the same. In other words, T should equal T_G. Who could possibly have imagined it to be otherwise? And so while Gertrude concludes that

SPEED OF LIGHT EQUALS 2L DIVIDED BY T

or, more abstractly,

SPEED OF LIGHT $= 2L / T$,

you will calculate that

SPEED OF LIGHT $=$ SOME NUMBER GREATER THAN 2L / T_G,

and since you assume that $T_G = T$, you calculate

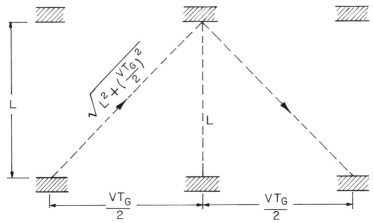

The Pythagorean theorem states that in a right triangle (a triangle in which one of the angles is "right" or 90°) the sum of the squares of the two sides is equal to the square of the hypotenuse:

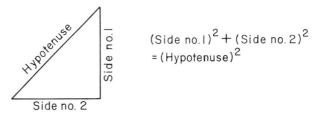

$$(\text{Side no.1})^2 + (\text{Side no. 2})^2$$
$$= (\text{Hypotenuse})^2$$

In the right triangle involving the light pulse, side no.1 = L, side no. 2 = $\frac{VT_G}{2}$ and so the hypotenuse (equal to the path followed by the light pulse as it moves between the bottom and top mirrors)

$$= \sqrt{L^2 + (\frac{VT_G}{2})^2}$$

FIGURE 3.15

SPEED OF LIGHT $=$ SOME NUMBER GREATER THAN 2L / T.

In other words, the value of the speed of light according to your calculations will be greater than the value calculated by Gertrude. You and Gertrude differ because although you both use the same time for the light's round trip in calculating speed, you observe a greater round trip distance for the light than she does.

For a Newtonian physicist this is fine. Why shouldn't you and Gertrude get different values for the speed of light? After all, you are moving through

Box 3.1
COMPARING YOUR CALCULATION OF THE
SPEED OF LIGHT WITH GERTRUDE'S

According to Gertrude: SPEED OF LIGHT $= 2L/T$
According to you: SPEED OF LIGHT $= [2\sqrt{L^2 + (vT_G/2)^2}]/T_G$

For a Newtonian scientist, $T = T_G$, and so,
According to Gertrude: SPEED OF LIGHT $= 2L/T_G$
According to you: SPEED OF LIGHT $= [2\sqrt{L^2 + (vT_G/2)^2}]/T_G$
$= [$A NUMBER GREATER THAN $2L]/T_G$
In other words, according to a Newtonian analysis you measure
a greater speed for light than does Gertrude.

---

space at one speed and she is moving at another, so light waves in space
should appear to move at one speed to you and at another speed to Gertrude. This is what a Newtonian physicist would have concluded, but it is
not what Albert Einstein concluded.

First, as we have seen, Einstein realized that you and Gertrude can never
agree on a common system of time; so for Einstein, the value of time that
you use in your formula for calculating the speed of light, T_G, need not
equal the value of time that Gertrude will use in her formula for the speed
of light. And then there is Einstein's second postulate from which he developed his relativity theory; this postulate says that the speed of light will
be the same no matter how the light source moves with respect to the observer.

Let us now analyze the observations of Gertrude's light clock as Einstein
would, in a radically non-Newtonian way. As we have already said, for a
Newtonian, Gertrude's system of time and yours will be one and the same,
and so you and Gertrude will measure two different values for the speed of
light. But Einstein, because of his careful analysis of the nature of time,
questioned the equivalence of Gertrude's time and yours; instead, he in-
sisted upon the validity of his second postulate and required that the two of
you get the same value for the speed of light. For this to be the case, it turns
out that the two of you must disagree on your systems of time.

You and Gertrude both calculate the speed of light by dividing the light
pulse's round-trip distance by the measured time required for that round

trip (see, again, optional box 3.1). But you determine that the light covers a greater round-trip distance than does Gertrude, and so for you to get the same speed, your measured time τ_G must be correspondingly greater than her time τ (or, to put this in slightly more mathematical terms, since the numerator in your formula for the speed of light is greater than Gertrude's, your denominator, the round-trip time, must also be greater if you are to get the same value for the speed of light). That is, Gertrude's reading of her clock and your measurements of the same clock must show different clock rates. If Gertrude says that her clock ticks once each second, you would be forced to disagree and say, "No, your clock is slow; actually I determine that it must tick at a greater time interval than once each second": τ_G must be greater than τ.

When Gertrude and her clock were at rest with respect to you, you both agreed that her clock ticked once each second; but now that the very same clock is moving with respect to you, you are forced to conclude that the clock must tick more slowly. Meanwhile, Gertrude determines that the clock continues to tick at its original rate. This discrepancy between Gertrude's measurement of her clock (the "proper time") and your measurement of her clock has been called "time dilation."

3.8 THE RELATIVITY OF LENGTH MEASUREMENTS

Having analyzed the meaning of simultaneity and the process of time measurement, let us use Einstein's approach to challenge still another notion of Newtonian physics. Once again, we can do this by addressing an operational definition: what exactly does it mean to measure the length of something? In particular, what does it mean to measure the length of something like Gertrude's car while it is in uniform motion with respect to someone standing by the railroad track?

The first thing to understand is that measurements of length are, ultimately, measurements of simultaneous events. This statement may be surprising at first, but it is entirely correct. Here is a familiar example: suppose you use a ruler to measure the length of something, say, the block of wood shown in figure 3.16. Measuring the length of the block when it is at rest with respect to the ruler is no problem at all. You place the ruler next to the block and observe the markings on the ruler that fall next to the two ends of the block. The block stays put so you can take all the time you need to compare the positions of its two ends with the two corresponding markings on the ruler. Even though it may take you quite some time to record these two

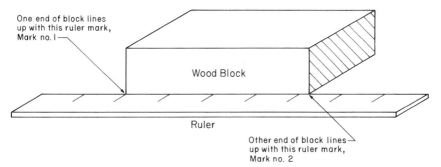

One end of block lines
up with this ruler mark,
Mark no. I

Wood Block

Ruler

Other end of block lines
up with this ruler mark,
Mark no. 2

Length of block is difference in two ruler readings : Length = Mark no. 2 – Mark no. I

FIGURE 3.16

ruler markings, you know that they apply for all the time the block is at rest. You know, in other words, the simultaneous positions of the ends of the block situated next to the markings on the ruler, even though you do not really read the two ruler markings simultaneously.

But if you now wish to use your ruler to measure the length of a moving object, like Gertrude's car, you have a problem. Imagine in figure 3.16 that the block of wood is sliding uniformly along the edge of the ruler. As the block moves past your location at the side of the ruler, you can instantaneously note the position of one edge of the block with respect to the markings on the ruler. But to get the length of the block, you need to know, *at that very instant*, where the other end falls along the ruler. In other words, you need to know the simultaneous positions of the two ends of the block with respect to the ruler markings. Once again, as in the operational definition of the common system of time, you need to determine two simultaneous events: the ruler markings that line up instantaneously with the two ends of the moving block.

Earlier, in discussing two simultaneous events, we dealt with flashbulbs fired at two different locations along the railroad track (see figure 3.5); we now deal with a situation that is fundamentally the same. The two events in question now are not the explosions of flashbulbs, but two alignments: the alignments of the ends of an object whose length we are trying to measure with the markings on a ruler. As with the two flashbulb explosions, these two alignment events occur at two different places in space: the two ends of the object whose length we are attempting to determine. In fact, it turns out that any length measurement of a moving object boils down to the de-

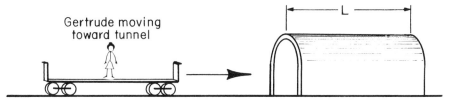

Gertrude moving
toward tunnel

L

FIGURE 3.17

termination of two simultaneous events. It is characteristic of Einstein's genius that he was able to recognize this fact.

As another example of this process, let us see how you might determine
the length of Gertrude's car as she moves past you on the railroad track (see
figure 3.17). Suppose there is a tunnel on the track whose length you have
carefully measured to be L. You stand exactly half way between the two
ends of the tunnel and you set up another mirror apparatus so that you can
view both ends of the tunnel at the same time. Gertrude moves uniformly
along the track at a constant speed v and enters the tunnel (figure 3.18).
You see the rear of her car vanish into the tunnel entrance at the same instant that the front of her car begins to leave the exit at the other end of the
tunnel. In other words, you conclude that Gertrude's car just fits in the
length of the tunnel. Since you have determined the length of the tunnel to
be L, you also report that as the length of Gertrude's car. Notice carefully
what lies at the heart of your length measurement of Gertrude's car: again,
it is the observation of two simultaneous events.

But how does Gertrude measure things? Remember the experiment we
discussed in connection with figure 3.8: we had two flashbulbs at different

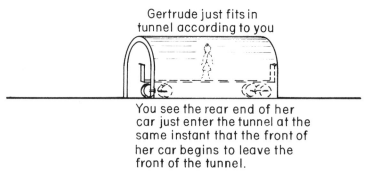

Gertrude just fits in
tunnel according to you

You see the rear end of her
car just enter the tunnel at the
same instant that the front of
her car begins to leave the
front of the tunnel.

FIGURE 3.18

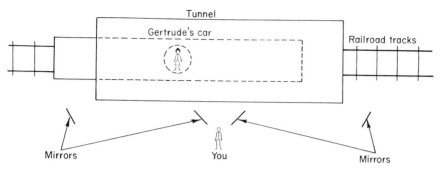

FIGURE 3.19

locations along the railroad track, and in that case Einstein showed that
flashes of the two bulbs (or, more generally, two events) appearing simultaneous to you will not appear simultaneous to Gertrude. Therefore, because length measurements are also determinations of simultaneous events,
Gertrude will not reach the same conclusion as you regarding the length of
her car. Let us look at this situation in greater detail. Figure 3.19 shows a
bird's-eye view of the tunnel with Gertrude's car just disappearing into it.
You have set up your system of mirrors, which permits you to view both
ends of the tunnel without moving your head. Gertrude has a similar apparatus to view both ends of her car without turning her head. At a certain
time, according to you, the front of Gertrude's car just begins to emerge
from the exit of the tunnel at the exact instant that the rear of her car van-
ishes into the tunnel entrance (figure 3.20). That is why you conclude that
Gertrude's length is equal to that of the tunnel. On figure 3.20, these two
events have been indicated as two flashes of light, just like the two flash-

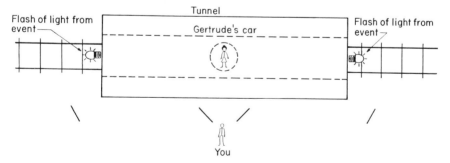

FIGURE 3.20

bulbs discussed earlier and shown in figure 3.8, only now the two flashes mark the alignment of the two ends of Gertrude's car with the two ends of the tunnel.

We know that eventually these two flashes of light are going to reach you at the same instant because we have supposed that they are simultaneous to you. But what does Gertrude see? Figure 3.21 is another bird's-eye view of the situation a moment later. Now the light from the two events is on its way both to you and to Gertrude, but before you can see either event Gertrude sees the light from the event marking the alignment of the front of her car with the tunnel exit; and so Gertrude concludes that at that instant the front of her car has just left the tunnel. Then, a little later on (figure 3.22), Gertrude sees the light from the event marking the alignment of the rear of her car with the tunnel entrance; and so she concludes that at that later time, the rear of her car is just disappearing into the entrance of the tunnel. Still later (figure 3.23), you see the light from both ends of Gertrude's car simultaneously, indicating to you that both ends of her car lined up with the ends of the tunnel simultaneously.

So, while you see the two alignments of the ends of the tunnel with the ends of Gertrude's car as simultaneous, Gertrude first sees the front of her car leave the tunnel exit, and then *later* she sees the light from the rear of her car entering the tunnel. That is, the signal indicating to Gertrude that the rear of her car has just entered the tunnel comes to her after she knows that the front of her car has already left the tunnel. Therefore, Gertrude concludes that her car must be longer than the tunnel. Or, to put matters in your perspective, you measure Gertrude's length to be shorter than she measures it to be. Her length appears to be contracted to you, and all because you and Gertrude cannot agree upon what events are simultaneous. This discrep-

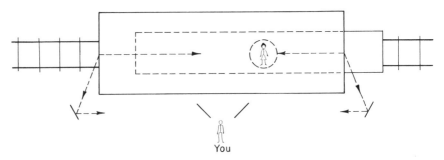

FIGURE 3.21

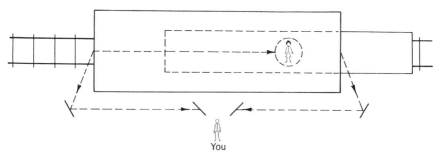

FIGURE 3.22

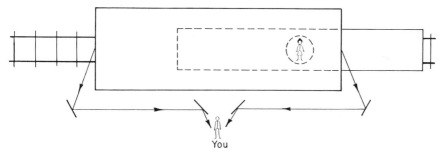

FIGURE 3.23

ancy between Gertrude's measurement of her own length (called the "proper length") and your measurement of that same length has been called "length contraction." Again, we emphasize that you are *both* correct in your respective length determinations within the context of your inertial frames of reference.

3.9 SUMMARY AND CONFIRMATION OF THE SPECIAL THEORY OF RELATIVITY

We have seen that all measurements of time and length are really measurements of simultaneous events. Because Einstein was able to show that you and any observer in motion with respect to you (like Gertrude) cannot agree on what you mean by simultaneous events, the two of you cannot synchronize your watches and you will be unable to agree on measurements of length.

All of this sounds very strange. By asking the most simple sorts of questions, by pursuing operational definitions for commonly used quantities without regard for the cherished Newtonian notions of space and time that he demolished along the way, Einstein succeeded in creating a profoundly non-Newtonian vision of the world, a vision that we may find disturbing. Nevertheless, Einstein's logic was so clear, his operational definitions seemed so sensible, that he had the courage to reject the established conception of physical reality and to adopt a new one. In particular, Newton's model of motion and the mechanical view of the world that went along with that model had to be revised. If the heliocentric world view described in chapter 1 demanded a "new physics," Einstein's theory demanded that a still "newer physics," a post-Newtonian physics, be created.

As the years passed since Einstein's paper of 1905, many of the bizarre predictions of special relativity have been tested repeatedly in the laboratory, and the theory has been amply confirmed. Consider the experiment conducted by the physicists Rossi and Hall in 1941, which confirmed that the measured rate at which a clock ticks will be different for observers at rest and in motion with respect to the clock.[18] Although the details of the actual measurements are rather indirect and far more complex than our brief description suggests, the idea is this. A particle called the mu-meson has a limited lifetime of a few millionths of a second. After it has existed for its lifetime, it decays by splitting itself into three other sorts of particles. The lifetime of an individual mu-meson may fall in a rather wide range of values (just as the lifetime of an individual human being may fall in a wide range of values) but the average lifetime figured for a large number of particles is well defined (just as the average lifetime of a human is well defined for a large population sample). The mu-mesons therefore represent a natural clock. They live for a certain (and well-known) average amount of time, and then they decay. The average lifetime of many mu-mesons at rest can be determined and compared with the measured average lifetime of mu-mesons moving at a very high speed (in the original experiments, the speeds were in excess of 99 percent of the speed of light). According to Einstein's relativity theory, the moving mu-mesons should appear to "live longer"—that is, their clocks should run more slowly than they do when at

[18] The original paper, written in technical language, is B. Rossi and D. B. Hall, "Variation of the Rate of Decay of Mesotrons with Momentum," *Physical Review* 59 (1941): 223–228. More recently, D. H. Frisch and J. H. Smith produced a film version of the same experiment for use in physics classes, "Time Dilation—an Experiment with Mu-Mesons," Education Development Center, Newton, Mass., 1963; see also David H. Frisch and James H. Smith, "Measurement of the Relativistic Time Dilation Using Mu-Mesons," *American Journal of Physics* 31 (1963): 342–355.

rest. This was found to be the case: the rapidly moving mu-mesons lived about nine times longer than similar particles at rest with respect to the observers, just the amount of time difference predicted by Einstein's theory for particles moving at that speed.

In other words, Einstein's vision of reality has been found to be superior to Newton's in its predictive power. When a clock is moved past an observer, the clock really does tick more slowly than when it is at rest with respect to the observer, and the amount of the slowing down is what relativity theory predicts it should be.

It is important to point out that these relativistic aspects of the physical world remained hidden from students of nature for millennia because they are ordinarily so small as to be negligible. For example, for a clock to appear to tick at a rate 15 percent slower than normal, it would have to move past an observer at half the speed of light (that is, it would have to be moving past at 93,000 miles a second). Very sophisticated instrumentation is required to measure such effects, and that instrumentation has only recently become available. Nevertheless, the relativity effects do turn out to be observed.

3.10 THE ROLE OF LIGHT

Why should the sorts of experiments with light rays that we have been talking about and a postulate based on the constancy of the speed of light dictate that moving clocks tick more slowly than stationary clocks, or that lengths that we measure for objects should depend upon how rapidly the objects move with respect to us? What is so special about light that it seems to hold the key to the discovery of all these strange aspects of space and time?

We retrace our steps for just a moment to see exactly what prompted our conclusion that moving clocks must tick at a different rate than stationary ones. In figures 3.13 and 3.14 we had a light clock set up in Gertrude's railroad car, and by timing the round trip for one pulse of light, both you and Gertrude were able to calculate the speed of light. But Gertrude saw the light cover less distance in one round trip than you did. Therefore, since the speed of the light is given by

$$\text{SPEED} = \text{ROUND TRIP DISTANCE} / \text{ROUND TRIP TIME},$$

we had to assume that your measured time was correspondingly greater than Gertrude's so that the two of you would get the same value for the speed of light, as demanded by Einstein's postulate that the speed of light is the same regardless of the state of motion of the source of light. Notice that this conclusion regarding the difference in your and Gertrude's time measurements has nothing whatever to do with the type of clock you are using. If the postulate about the invariant character of light's speed is to be fulfilled, your time measurements, however they are accomplished, must differ from Gertrude's. And, as we have pointed out, this prediction, and all of the others that result from the insistence on the validity of Einstein's second postulate, are amply confirmed—confirmed in experiments not only involving light, but experiments with many other things too, like particles in cyclotrons, naturally occurring cosmic rays, and atomic clocks. How does light and its adherence to Einstein's postulate connect with these diverse phenomena? And there is a related question. Why is the speed of light a constant for all inertial observers regardless of their state of motion relative to the light source? Is this something to be explained, or must we simply accept it as "a fact of nature?" It may be that understanding this strange property of light will result in greater insight into the connection between the phenomenon of light and measurements of time and space. As it is, we must leave open the questions raised by this connection.

3.11 SPACE, TIME, AND SPACETIME

Our presentation of Einstein's special theory of relativity per se is complete. However, we wish to close this chapter with a discussion of an important way of looking at Einstein's creation that developed in the aftermath of the 1905 paper. In fact, this later approach to relativity was adopted by Einstein as he developed his general theory (discussed in chapter 5). The mathematician and physicist Hermann Minkowski (1864–1909) articulated this new development in an address in 1908:

> The views of space and time which I wish to lay before you have sprung from the soil of experimental physics, and therein lies their strength. They are radical. Henceforth space by itself, and time by itself, are doomed to fade away into mere shadows, and only a kind of union of the two will preserve an independent reality.[19]

This is the germ of the spacetime idea, and Minkowski developed it into a

---

[19] H. Minkowski, "Space and Time," translation reprinted in Lorentz et al., *The Principle of Relativity*, p. 75; this address was not the first announcement of Minkowski's approach. See Clark, *Einstein*, pp. 157–160, and Pais, *Subtle Is the Lord*, pp. 151–152.

detailed, mathematical description of relativity theory. It is important that we explain just what is meant by Minkowski's remarks, and how this view is used by physicists, because it is sometimes misunderstood to mean that space and time are somehow similar things, or even manifestations of the same essence. Here our intuitive sense of the way that space and time operate in the world appears to guide us well. Space and time are very different, measured in very different ways (as our discussions in sections 3.7 and 3.8 indicate) and sensed differently by us.

Above all, Minkowski's mathematical way of stating the theory of relativity is simple and useful. It provides physicists with a highly pictorial way of dealing with relativity in quantitative terms. Indeed, we too will find this pictorial representation useful, particularly in our discussion of general relativity in chapter 5.

Let us begin by discussing a bit of necessary technical terminology. If we wish to locate a point in space quantitatively, we must make distance measurements from some arbitrary reference point. For example, on our railroad track we could locate some zero-mile stake and measure Gertrude's and our positions with respect to that. Because all motion is confined to the track (one direction in space), we need only one number to specify the position of things: the number corresponding to distance measurements from our reference point. The units of the distance measurements are completely irrelevant. One number in any system of units will do. This number, required for specification of an object's location in space, is called a "dimension," and because only one number is needed along the track, we say that the track represents a "one-dimensional system."

In general, objects are not constrained to move along a rigid track. Suppose that we wish to specify (quantitatively) the location of an object anywhere on the surface of the earth. Geographers do this by means of a grid of longitude and latitude that they imagine to cover the earth's surface. Stating the longitude and latitude numbers for a location on earth is all that is needed to identify the position uniquely. Each of the two numbers needed to locate the position of something on a surface is also called a "dimension," and a surface represents a "two-dimensional system" because two location numbers are needed.

But now suppose that we need to locate something not bound to the surface of the earth—for example, a bird in the air or a thrown baseball. Then we need three numbers (dimensions) to specify location: we can still use longitude and latitude to specify the point on the ground directly below the object, and to these we add a third number specifying the height of the object above that point on the ground. Experience shows that three numbers are the maximum ever needed to specify the location of something in space. Accordingly, we can say that we live in a "three-dimensional space."

Minkowski went one step farther. In relativity we deal not just with the location of *objects* in space, but with happenings or *events* that are located in space and in time. To specify an event completely we must not only say what happened (a flashbulb fired or the front of a train left a tunnel), we must also say where it happened and when it happened. The "where" is taken care of by our three dimensions of space. The "when" requires a fourth number, a clock reading.

So far, we have really added nothing to what Einstein said in 1905. His development of relativity recognized the necessity of stating four numbers to specify an event in space and time. What Minkowski developed was a mathematical (not physical) unification of these four numbers designating dimensions. Instead of thinking of three spatial dimensions and one time dimension, Minkowski proposed thinking of the world as a grid of four-dimensional *spacetime*.

The primary advantage to this shift in view was a substantial degree of simplification in the mathematical treatment of relativity, a process that will not concern us. But we do want to point out the pictorial or graphical counterpart to this mathematical simplification, something called a "space-time diagram" or "Minkowski diagram."

We begin by describing things on our railroad track again; in fact, let's go back to the very first picture of the track, figure 3.1. Here we see a flash-bulb and two observers. But without the words that went along with this picture, what does it tell us? Only that there are a track, a flashbulb, and two observers. Yet the diagram was meant to illustrate a very important point about the time sequence of events, namely, the flashing of the bulb and the subsequent detection of the flash by observers at points A and B. Using Minkowski's technique of picturing events in spacetime, the picture can be made to represent the events and their temporal relationship with no additional words.

In figure 3.24 we have reproduced the original drawing of the railroad track shown in figure 3.1; we see, as before, the bulb and the two observers, now identified as Pablo and Alice. And notice that we have added a reference marker on the track. This is a position from which we agree to measure all locations of things and events along the track. Directly above the track we have drawn a table of position and common time for three events. We measure the positions of the events on the actual track (using

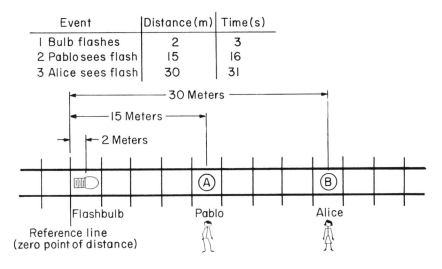

Event	Distance (m)	Time (s)
1 Bulb flashes	2	3
2 Pablo sees flash	15	16
3 Alice sees flash	30	31

FIGURE 3.24

whatever kind of ruler we wish), and then we enter those values in the table; next, we record the times for the three events according to the common system of time that we established along the track and as determined by clock readings at the bulb and at Pablo's and Gertrude's positions. (The time and distance values shown in the table and on the picture of the track in figure 3.24 are entirely arbitrary.)

We are now ready to draw the spacetime (or Minkowski) diagram for these events (figure 3.25). We have two perpendicular lines called "axes." The horizontal line is a distance or space axis. We locate the distance of each event along the scale marked on this axis and draw a dashed line through this distance point perpendicular to the space axis. Next we locate the time values for each event along the scale marked on the common time axis. We draw dashed lines through these points perpendicular to the time axis. Where the dashed lines from the time and distance measurements of event 1 meet we draw a point; similarly where the dashed distance and time lines for events 2 and 3 meet we draw points. The result is a Minkowski diagram for the three events. Points in a Minkowski diagram represent events (that is, position and time measurements from a timetable).

Why bother with all of this? Mainly because we can now see the temporal relation of the three events without reading through a paragraph. The three events are arrayed in distance (space), as represented on the horizon-

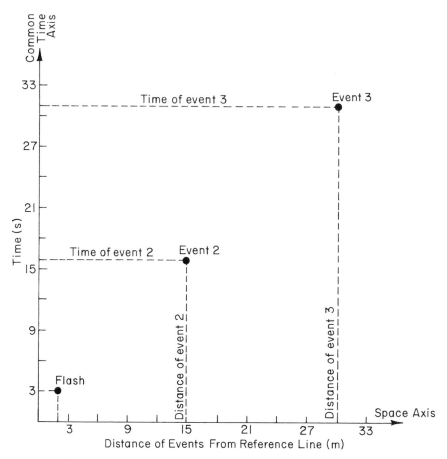

FIGURE 3.25

tal axis, as well as in time, represented along the vertical axis, and we can quickly see how they stand in relation to one another. But there is more. Turn back to figure 3.7. This was our way of representing the fact that Gertrude moves along the track. It shows only that Gertrude is on the track (a spatial relationship) and the arrow is an abstract way of suggesting her motion. But we don't know what *sort* of motion she is undergoing. Is it uniform or is it changing? And how fast is her speed in miles per hour? From the picture alone we cannot tell. Again we must read the paragraphs that went with the picture. Or consider the even more complex sequence of events depicted in the series of figures 3.19 through 3.23. Here the situation was sufficiently complex that we had to create what amounted to a series of "stills" from a "movie" of Gertrude's motion through the tunnel.

But Minkowski's approach can describe the pertinent features of the whole "movie" with just one picture. The spacetime diagram in figure 3.26 is drawn for the observations that you would make at rest by the side of the track (the distances and times are those you would measure) and it provides detailed information about Gertrude and the two ends of the tunnel. First, there is a line indicating the position of the front of Gertrude's car at each instant of time. A line like this in a Minkowski diagram showing how an object moves along in space and time (that is, how it moves through spacetime) is called a "world line." Points along this world line correspond to position and time readings that you would make for the front of Gertrude's car as it moves along. Notice the great economy of this picture. We don't need a vast timetable giving Gertrude's location at each instant of

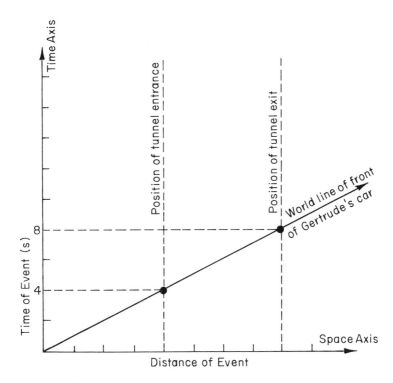

FIGURE 3.26

time. We can show it all in a much more simple diagram. Should we want numbers from the diagram, all we do is locate the position for which we want to know the arrival time of the train front, and draw a vertical line through this distance value. We make a mark at the point where this line intersects Gertrude's world line. Through this point we draw a horizontal line over to the time axis and read the time value. That will give us the time number we want.

Notice that the positions of the tunnel entrance and exit are marked by dashed, vertical lines. Because the tunnel is at rest with respect to you, the position of the tunnel does not change with time and so these world lines are vertical. Where the train's world line and the world line of the tunnel entrance cross we have an event: at one value of time the train front and the tunnel entrance have the same position value in space. In other words, the time of this event is the time at which the front of Gertrude's car reaches the tunnel entrance. From the diagram we see that this happens at 4 seconds (again, the numbers used in this diagram are arbitrary); similarly we find that the front of the train reaches the tunnel exit at 8 seconds on the common time system. We have reduced our "movie" to a single diagram, with the motion embodied in the world lines of the various objects. We can even tell that the motion involved is uniform, because uniform motion is represented by straight world lines in Minkowski diagrams. Were Gertrude's train to accelerate, we would see her world line bend.

Figure 3.27 shows world lines for both the front and the rear of Gertrude's car as it moves along the track. Again, remember that these diagrams refer to your measurements and not to Gertrude's. Notice that you see the rear of her car enter the tunnel at the exact instant (8 seconds) that the front of the car leaves the tunnel; that was your reason for concluding that Gertrude's car has the same length as the tunnel.

It is worth mentioning another advantage to the use of these spacetime diagrams. All of the diagrams we have drawn so far are based on your observations of things. As we know, Gertrude will reach different conclusions from yours regarding time and distance measurements—that is, she will draw different spacetime diagrams for the same events. But it turns out that her diagrams may be easily obtained from yours using graphical techniques. We will not go into details here (they are included in appendix A); suffice it to say that this one diagram, drawn from the perspective of one observer, may be used to obtain easily the time and distance measurements that any other inertial observer would make.

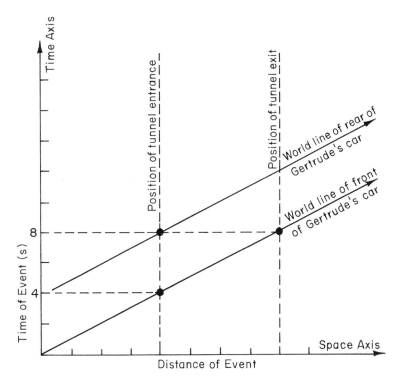

FIGURE 3.27

This discussion of Minkowski diagrams brings to sharp focus a point worth emphasizing. Relativity theory not only predicts that different inertial observers will disagree on the results of various sorts of physical measurements, it also predicts by exactly how much various observers will disagree. Thus each inertial observer can use Einstein's model to calculate exactly what any other inertial observer will measure. The Minkowski diagram provides a graphical way of carrying out this calculation.

One more point. We have shown diagrams pertaining to spatial arrangements in just one dimension. How do spacetime diagrams look in situations not confined to a straight railroad track? Figure 3.28 is a sketch of an ice-skating rink with a closed curve etched on its surface by an (invisible) ice skater. Since her motion is confined to the surface of the rink, two numbers will serve to locate her position on the ice; that is, we are dealing here with a two-dimensional situation in space. Again, our diagram lacks much de-

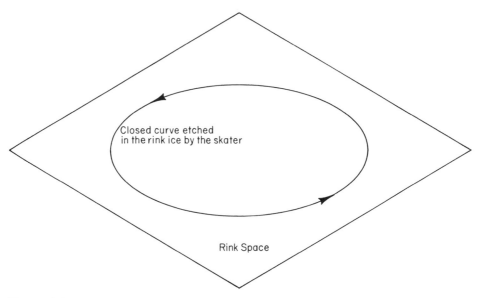

Closed curve etched
in the rink ice by the skater

Rink Space

FIGURE 3.28

tail. From it alone we cannot tell her speed (or whether or not it is constant) or whether she completes only one circle or retraces the same path over and over. We need words and a timetable to fill in the details. Let us look at the same events in a full spacetime diagram (figure 3.29). Again, we see the plane of the rink, but now at one corner we have erected a vertical time axis. As with our previous spacetime diagrams, we locate the position of events in space and in time; here we use the vertical time axis as we used the previous vertical axis when we dealt with one-dimensional spatial situations. The skater's world line is now curved (she is accelerating), and it takes on the form of a corkscrew or helix, moving upward in the diagram.

Notice that for a one-dimensional spatial situation we needed two dimensions (a plane) to draw our spacetime diagram because when we plot time we add another dimension to our description of things. In the case of the skater, we started with two spatial dimensions and our spacetime diagram became three-dimensional (and our representation on the page of this book became an abstraction from what is a fully three-dimensional thing). Were we to make a spacetime diagram for events taking place in three dimensions of space, we would be required to draw a four-dimensional diagram. We cannot do that. In fact, it is even hard to imagine such a thing. Fortunately, the mathematics that goes along with these diagrams does all the hard

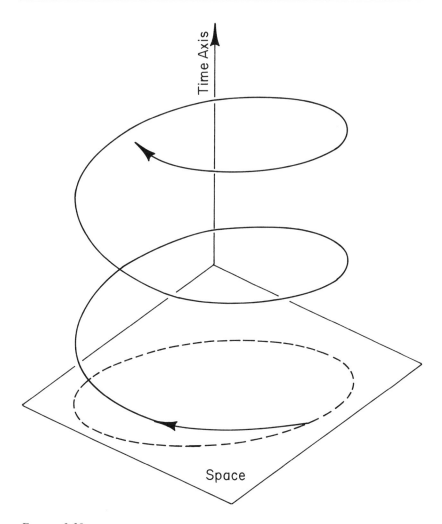

FIGURE 3.29

work, so that in practice physicists need not imagine 4-dimensional diagrams. Minkowski diagrams in two and three dimensions are, however,
useful in visualizing one- and two-dimensional spatial situations, and quite
often the mental pictures that develop in these limited cases help in analyzing fully four-dimensional descriptions of spacetime.

4

SOME CONSEQUENCES OF
SPECIAL RELATIVITY

> Who knows, perhaps He is a little malicious.
>
> —A. Einstein

4.1 Introduction

In the previous chapter we discussed the basic ideas behind Einstein's 1905 theory of relativity, and we indicated how measurements of time and distance would differ for observers moving relative to each other. We now consider some of its consequences for several physical situations. Again our discussion will not be mathematical; as a consequence some of our arguments may seem incomplete, and certain results appear without full elaboration (although we will indicate references to more complete treatments in our appendixes or in other works).

4.2 The Lorentz Transformations

While concluding that different inertial observers will obtain different results in measuring times and positions for events, Einstein's 1905 paper also provides everything necessary to calculate just what those differing results will be. This quantitative apparatus is embodied in two equations called the "Lorentz transformations" (box 4.1, optional).[1] These equations permit one to convert distance and time measurements of events made by any inertial observer to measurements that would be made by any other inertial observer. The Lorentz transformation equations therefore summarize the quantitative results of Einstein's theory. Although we will not be concerned with the manipulation of these equations, they have one impor-

---

[1] See also section 2.7.

Box 4.1
THE LORENTZ TRANSFORMATIONS

For an observer at rest by the tracks, D and T are the distance and time measured for an event.

For a second observer riding in the railway car at a speed v with respect to the first observer, D' and T' are the distance and time measured *for the same event*.

Einstein showed that:

$$D' = \frac{D - VT}{\sqrt{1 - \dfrac{V^2}{C^2}}} \quad \text{and} \quad T' = \frac{t - \dfrac{VD}{C^2}}{\sqrt{1 - \dfrac{V^2}{C^2}}},$$

where c is the speed of light.

These two equations are called the Lorentz transformations. Given T and D or T' and D', the other two measurements can be calculated.

---

tant feature that we must note. The expression for T', the time of occurrence that Gertrude measures for an event on her common time system, involves not only T, your measured time for the event, but D, your measured position for the event. In other words, *when* Gertrude measures the event is related in these equations to *when and where* you measure it. Similarly, the value for the position that Gertrude measures for the event, D', is related to both *where and when* you measure it.

Distance and time measurements interweave in calculating Gertrude's measurements from yours or vice versa. Here again Minkowski's portrayal of spacetime can help in making this interrelationship more clear, but our caution in section 3.11 about overinterpreting the notion of spacetime bears repeating. Space and time measurements are interrelated in the Lorentz transformations to transpose measurements made by one observer into those of another observer, but the measurements of space and those of time are completely distinct. Space and time do not get "confused" with one another or "interchanged" with one another as is sometimes suggested in connection with relativity theory.

In order to comprehend the essence of the Lorentz transformations, imagine Gertrude riding again along the railroad track used so often in the last chapter; as usual you are standing on the ground beside the track. For any given value of Gertrude's speed with respect to the track, the Lorentz transformations permit us to calculate the rate at which you see Gertrude's clock tick. We carry out this calculation for various values of Gertrude's speed from zero (when she is at rest with respect to you) to nearly the speed of light. The results of these calculations represent the meaning of the Lorentz transformation of time measurements, and the calculated numbers could be presented in a table of Gertrude's speed and her corresponding clock rate. But it will be more useful to see these results graphically. In figure 4.1 the vertical direction on the graph corresponds to the number of ticks that your clock makes for each single tick that you determine Gertrude's clock makes as it moves past you. The horizontal direction represents Gertrude's speed along the tracks as a percentage of the speed of light.[2] The points on this graph corresponding to each calculated value of clock rate have been connected by a solid black line. This line represents the Lorentz time transformation.

We can use figure 4.1 in the following way. Suppose we want to know at what speed you make the measurement that Gertrude's clock ticks three times slower than she determines that it ticks. In this case you will measure three ticks on your clock for every one that you determine Gertrude's clock makes, and so we locate the number 3 on the vertical (time) scale of the graph. Next we draw a horizontal dashed line from the mark corresponding to three ticks to the solid curve which represents the Lorentz transformation. This dashed line intersects the Lorentz transformation curve at the point P shown. From this point we draw a vertical dashed line to the speed scale on the graph. This vertical line intersects the speed scale at a value of 94.3 percent, which gives the desired result: the speed—in percent of the speed of light—required for your measurement of Gertrude's clock to be three times as slow as Gertrude's measurement of her own clock. Of course

[2] In figure 4.1 (and in other graphs of relativity effects in this chapter) we plot ratios or percentages of quantities (percentages are really ratios multiplied by one hundred). There are two reasons for this. First, a ratio of two quantities, such as a ratio of speeds or of times, avoids the utilization of specific units of measurement such as hours, seconds, miles per hour, or kilometers per second. The graphs of ratios will be valid whatever system of units one wishes to think about. Second, the speed of light is a huge number in any ordinary system of units: 186,000 miles per second or 300,000 kilometers per second. Graphs not using speed as a percentage of the speed of light would have to be labeled with large numbers—cumbersome to read and to interpret.

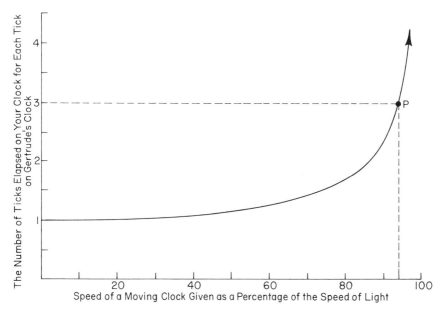

FIGURE 4.1

the problem can be worked the other way around. We can ask how many ticks you determine that your clock makes for each tick of Gertrude's clock when she moves past you at 94.3 percent of the speed of light. The same two dashed lines may be drawn to find the result.

We may represent the essence of the other Lorentz transformation in a similar manner (figure 4.2). As Gertrude moves along the track, we know that you will measure a different value for the length of her car than will Gertrude. Given Gertrude's speed with respect to you, the second Lorentz transformation can be used to calculate the length that you measure. Along the vertical scale in figure 4.2 we plot your measured length of Gertrude's car as a percentage of Gertrude's own measurement of her car. As in figure 4.1, the horizontal scale gives Gertrude's speed along the track as a percentage of the speed of light. A construction of horizontal and vertical dashed lines similar to that shown in figure 4.1 has been drawn in figure 4.2 to indicate the speed at which you measure Gertrude's length to be half (50 percent) of the value she measures. That speed is 86.6 percent of the speed of light.

From the last two figures we see in quantitative terms something we discussed qualitatively in the last chapter. When Gertrude is at rest with you

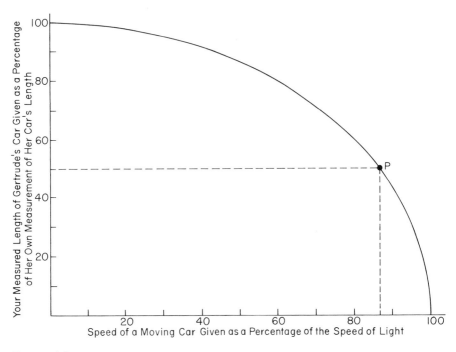

FIGURE 4.2

on the track (her speed is zero, or zero percent of the speed of light) her length and time measurements are identical to yours. When Gertrude begins to move on the tracks, her clock appears to you to tick more slowly (the number of ticks on your clock between ticks on hers increases in figure 4.1), and her length appears to shorten (the length percentage decreases in figure 4.2). These graphs were calculated with care and you may use them to read actual values of such effects.

4.3 Not Everything Is Relative

It is sometimes said that according to Einstein's theory of relativity "everything is relative." Certainly the Lorentz transformations do predict that measurements of time and distance for events will differ for different observers, depending upon their positions in space and their relative states of motion. One can then say with accuracy that time and distance (or length) measurements are specific to an observer's position and state of motion; they are in that sense "relative" to each observer. Relativity theory can be used to show that measurements of other quantities used to describe the physical world (like energy and mass, see section 4.5) depend upon an observer's state of motion and position. But it is incorrect to state that "everything is relative" according to Einstein's theory. On the contrary, the theory mandates that some quantities are the same for all inertial observers regardless of their position or state of motion. Physicists refer to these constant quantities as "invariants."

The constant (invariant) character of the speed of light is a postulate of Einstein's 1905 theory, as we pointed out in section 3.6. Any inertial observer must measure the same speed for light. Since the speed for light is defined as the distance traveled by a light flash divided by the time required to travel that distance, the constancy of light speed imposes a constraint of sorts on time and distance measurements of events involving light flashes. To make that constraint clear, consider two events involving the same flash of light. Again we will use the railroad track with you and Pablo standing at different points along its side. The first event is the passage of a flash of light through your position by the track; the second event is the passage of the same light flash by Pablo. Different inertial observers (for example, you at rest by the track and Gertrude moving uniformly along the track in her car) will measure different time and distance values for these two events marking the progress of the light flash. But those measurements made by different inertial observers must always follow this rule: the measured distance traveled by the light flash divided by the measured time for the flash to cover that distance always equals the same value for the speed of light. Gertrude might choose to move past you and Pablo at a faster speed so that she determines a smaller and smaller distance separating you and Pablo, but her measured time difference between the two events must always be such that she calculates the same speed for light. In other words, Gertrude is not free to choose her state of motion so that any combination of distance and time differences for these events is observed, only those combinations of time and distance measurements that will result in her calculating the same speed for light.

The constraint on Gertrude's time and distance measurements of the two events is expressed by physicists in terms of a number called the *interval*, a number easily calculated by any observer from her time and distance measurements.[3] Relativity theory shows that this calculated interval be-

---

[3] Again we wish to caution readers not to confuse the colloquial meaning of a word (the word "interval" in this case) with its meaning in physical science. The physicist's interval

tween the two events involving you, Pablo, and the flash of light is an invariant. This means that no matter how Gertrude moves, her time and distance measurements of the events will always yield the same calculated interval. But there is more. Because the interval is an invariant, any inertial observer will calculate the same value as Gertrude does for the interval between these two events. So, to repeat, while Gertrude will measure different time and distance separations for two events depending upon her state of motion with respect to the track, the interval that she calculates from these time and distance separations will always be the same; furthermore, the number she calculates will be the same as that calculated by any other inertial observer of these two events.

The two events that we have been discussing so far are of a very special sort. They involve the passage of a light flash between you and Pablo. One can use the Lorentz transformations, however, to show that the calculated interval between *any* two events in spacetime will have the same value for all inertial observers (that is, the *interval* is invariant for all events, not just those involving light flashes). Consider, for example, two general events in four-dimensional spacetime. The first event is marked by Pablo applying his signature to a sketch of the shops along Montmartre as he sits at a café on the sidewalk. The second event is marked by Alice laughing at the sketch after Pablo walks over to her table some moments later. Different inertial observers will measure different times and positions for these two events, but they must always calculate the same value for the *interval* between them.

To summarize this discussion, according to the 1905 theory there indeed are certain aspects of the physical world (time and space separations of events) that are "relative." We understand the term "relative" to mean that if we follow well-defined protocols for measuring these aspects of the world, then the measurements depend upon the particular situation of the observer (in particular on the observer's state of motion). But there are also other aspects of the world—such as the interval or the measured speed of light—that are "constant" or "invariant," meaning that they are independent of the observer's situation. The theory of relativity could as easily be called "the theory of invariants" because in calculations using the theory, physicists often focus attention on the invariant quantities precisely because they are situation-independent.

---

is defined by an equation involving the spatial and temporal specification of two events in spacetime. The interval in spacetime is discussed and defined quantitatively in appendix A.4.

4.4 THE "ULTIMATE SPEED LIMIT"

Figures 4.1 and 4.2 show that as Gertrude's speed becomes greater, according to you her length becomes smaller and her clock ticks more slowly. There is a limit implied by these trends. If Gertrude were to move past you rapidly enough, you would determine that her length becomes vanishingly small. Similarly, if she were to go past you rapidly enough, you would determine that her clock had come to a stop. Inspection of the graphs shows that both of these extreme consequences occur at the same speed: a speed equal to 100 percent of the speed of light.

The very idea that there is some upper limit to the speed with which Gertrude can move along the tracks with respect to you is counterintuitive in our Newtonian world view. It is easy to imagine—and completely consistent with Newtonian physics—that given a sufficiently powerful engine with enough fuel, we should be able to accelerate Gertrude in her train to any speed with respect to the tracks. But according to special relativity theory such acceleration is not possible.

Although the imposition of a maximum ultimate speed may seem counterintuitive, we now hasten to point out that an *absence* of an upper speed limit (that is, our usual Newtonian view) can lead equally to counterintuitive situations.[4] Suppose, for example, that there were no maximum speed. Suppose, in fact, that it were possible to move Gertrude around at infinite speed with respect to you. In that case, Gertrude could move anywhere in the entire universe instantaneously. She could be ubiquitous.

Furthermore, we can argue a special role for light in this connection (again we will use the term "light" to stand for any form of electromagnetic radiation). Remember that light is a medium for the transfer of information: images received by the eye or signals coded in some way for reception by a telecommunication device. Consider a televised image broadcast out into space. If we were able to travel faster than the speed of this signal, we could "run ahead" of the signal and, by turning around to receive it, "replay" the event at any time. Indeed, we could now send such superfast messengers far out into space to catch the light sent out by the storming of the Bastille, or the Sermon on the Mount, or the formation of the earth and sun. By moving toward or away from the source of the signals at various speeds, we could observe events rapidly forward or backward, rather like a film run in either direction (also with the capability of "freezing the

---

[4] For ideas presented in the remainder of this section, we are grateful to Professor of Physics Joseph D. Harris of Dartmouth College.

frame'' by moving along with the light from the event of interest). We could even see ''effects'' take place before their ''causes.'' This is beginning to sound very strange. But there is more.

We show in appendix C that according to the Lorentz transformations, if it were possible to move things around at a speed greater than that of light (and it is not at all clear that this is possible), effects might actually precede their causes. We are not speaking here of passively observing events in reverse order like our superfast television viewer of the last paragraph. Here we are speaking of the direct experience of events preceding their causes. This sounds obscure without a specific example. Here is the situation analyzed in detail in appendix C. You and Gertrude agree to what amounts to a ''supraluminal'' tennis match (using faster-than-light tennis balls) as she moves past you on the railroad track at half the speed of light. You have all of the tennis balls, and you agree to serve to Gertrude; she will of course return your serve just as soon as she receives it. But with faster-than-light tennis balls, things can be arranged so that you receive Gertrude's return before you serve to her in the first place. You can experience ''the effect'' (the return of the ball) before ''the cause'' (your serve to Gertrude). And suppose you really did experience Gertrude's return before you served and then you decided *not* to serve to her. How could you receive a return if you decided never to serve?

The point is that motion faster than light would lead to grave consequences for our sense of causality and thus of ''reason.'' Therefore, we have some interest in theorizing that the speed of light really does represent an unattainable value as Einstein's model predicts, despite this notion's counterintuitive quality in our Newtonian world view. On the other hand, our arguments have in no way proven that the speed of light is the ultimate speed limit. Our intent has been merely to show that some such speed limit is not unanticipated by our Newtonian intuition.

4.5 How Inertia (Mass) Varies with Speed

In our discussion of Newton's second law in section 2.5 we described the ''inertia'' of an object as a measure of the object's reluctance to undergo any change in motion. With the development of his relativity theory and the recognition that measurements of space and time cannot be interpreted as simply as had been assumed previously, Einstein went on to investigate the consequences of his work for Newtonian mechanics. In particular, in a second relativity paper published in 1905[5] Einstein was able to show that, in his words, "If a body gives off energy . . . its mass diminishes," and later he adds, "The mass of a body is a measure of its energy-content." Expressed quantitatively, this surprising result can be written in the famous equation

$$E = MC^2,$$

where E is the energy possessed by an object, M is its mass (the quantitative measure of inertia), and C is the speed of light. This result related two physical attributes of matter that, before Einstein's work, had seemed quite distinct: mass (the measure of inertia) and energy (defined by physicists as the capacity to do work).

We have stated this result without trying to show how it obtains from the Lorentz transformations because this demonstration would require far more mathematical detail than we wish to impose on our readers.[6] However, we can cite some experimental proof of its validity. In this experiment, energy is delivered to identical particles (they happened to be electrons), and in response to this added energy the particles move at various speeds; in fact, the observed speed of a particle may be used as a measure of its energy. While the particles move at this speed, a known force is exerted on them to measure their reluctance to change their state of motion. From the value of the applied force and the observed response to that force, one can calculate the mass (inertia) of the particles (using Newton's second law). Figure 4.3 summarizes the results of such an experiment: the measured speed of a particle (given in the figure as a percentage of the speed of light) has been plotted versus the measured inertia of that same particle (given as a percentage of the mass measured when the particle is at rest). The measured mass increases as the particle moves more rapidly and hence gains energy; this mass increase is just what special relativity predicts it should be.[7]

[5] A. Einstein, "Ist die Trägheit eines Körpers von seinem Energieinhalt abhängig?" *Annalen der Physik* 18 (1905); this appears translated as "Does the Inertia of a Body Depend upon Its Energy Content?" in *The Principle of Relativity*, pp. 69–71.

[6] A more detailed discussion may be found in A. Einstein, *Relativity*, pp. 44–48. This still does not present a derivation of the relation. For that one must go to a full mathematical treatment such as the one found in French, *Special Relativity*, pp. 20–22 and 167–174.

[7] Actual data for this sort of experiment may be found in French, *Special Relativity*, p. 23. The curve shown in figure 4.3 is essentially the same as that in figure 4.1, corresponding to the fact that mass measurements depend on relative speed in the same way as the measured rates of clocks.

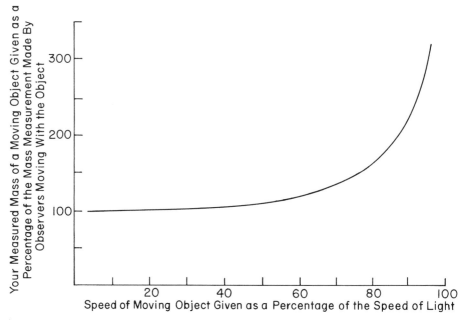

FIGURE 4.3

Special relativity theory also predicts that at the speed of light, an object would have infinitely large inertia. We can now develop the assertion made in the last section that it is impossible in practice to accelerate an object to the speed of light. Suppose we try it. Gertrude is on her railway car with a powerful rocket engine attached. She starts from rest, and as the rocket fires it applies a force. Gertrude's mass measures her reluctance to respond to the force of the rocket. Because the train starts out at very low speeds, Newton's laws hold to a very good approximation, so we can use Newton's second law ($F = M \times A$) to predict how fast Gertrude will move as a result of the rocket. But as the rocket accelerates Gertrude we know that her inertia will also increase. In other words, as she moves faster she offers more and more resistance to any further increase in speed. As her speed approaches that of light, her resistance to additional increases in speed becomes infinite and she simply cannot be pushed up to the speed of light. By expending enough rocket fuel (energy) we can make her approach the speed of light as closely as we wish, but she can never quite reach it.

So it is impossible to accelerate masses to the speed of light. But here we must make a careful distinction. It is not correct to say that nothing can move at the speed of light. After all, light moves at the speed of light, and so do other things such as the subatomic particles called neutrinos. These entities move at the speed of light because they were brought into existence moving at that speed (that is, acceleration was not necessary); in addition, they have zero inertia or mass.[8] Similarly, it is consistent with special relativity that entities could be brought into existence moving at a speed greater than that of light. Special relativity predicts that it is impossible to accelerate any matter *up to* the speed of light, but if the matter is brought into existence already moving at some speed greater than that of light, there is no problem—at least so far as consistency with special relativity theory is concerned. Such hypothetical particles (dubbed ''tachyons'') do create extremely serious problems in other ways, however, as our discussion of the ''supraluminal'' tennis balls in the preceding section indicated. Tachyons are discussed in appendix C.

4.6 SOME APPARENT PARADOXES RESULTING FROM APPLICATION OF THE SPECIAL THEORY OF RELATIVITY

A number of puzzles have occurred to students of relativity over the years, some of them resulting in a lively debate in the literature of physics. These are often phrased as statements of an apparent paradox. We will discuss three of these to illustrate both applications and misapplications of relativity theory.

A Scheme to Violate the Ultimate Speed Limit

We have said that in relativity theory the speed of light is an ultimate speed limit for any object, although any speed less than that of light is allowed. But schemes have been devised in attempts to violate this speed limit. Suppose we have two cars on our railroad track, Gertrude's and a new car conducted by Pablo. Both Gertrude and Pablo have powerful rockets to move their cars at three-fourths the speed of light with respect to you (at rest as usual by the side of the tracks). Such a speed is very high, but is perfectly allowable according to relativity theory. Suppose, then, Gertrude starts out on the track moving at three-fourths the speed of light to the right,

---

[8] There are as yet unconfirmed reports that neutrinos may have some very small mass; if this is so, according to relativity theory they cannot move at the speed of light but at some slightly smaller speed (the slight difference having been undetectable), and they are not brought into existence at the speed of light.

and Pablo does the same, except he moves to the left (figure 4.4). According to you, each car moves at three-fourths the speed of light but in opposite directions. What does Gertrude determine? If Newton's physics held, Gertrude should conclude that you and the tracks are receding from her at three-fourths the speed of light; she would also determine that Pablo recedes at three-fourths of the speed of light with respect to the tracks; but since the tracks themselves recede from her at three-fourths of the speed of light, she should measure Pablo receding from her at

THREE-FOURTHS PLUS THREE-FOURTHS =
1.5 TIMES THE SPEED OF LIGHT.

That is how speeds are combined in Newton's model: one adds the speeds together. As a result, Gertrude determines that Pablo is in violation of the ultimate speed limit.

But special relativity theory (and in particular the Lorentz transformation equations) dictates a different rule for adding speeds, and this relativistic rule predicts that Gertrude must measure Pablo moving at less than the speed of light. We can understand this result without appealing directly to the Lorentz transformations. Your measurements show that Pablo moves over a certain length of track in a certain amount of time so you determine that his speed is three-fourths that of light. From Gertrude's frame of reference, however, the distance that you conclude Pablo covers is contracted to a smaller value than you measure, and the time required for Pablo to cover this distance is greater than you measure (because to Gertrude your clock is slow). So Gertrude must conclude that Pablo's speed with respect to the track is less than you claim it to be. The faster Gertrude moves, the slower Pablo is measured by Gertrude to move with respect to the tracks; and so the resolution of the "paradox" can be understood qualitatively as a combination of length contraction and time dilation. Box 4.2 (optional) shows the correct formula to use in combining speeds according to the special theory of relativity. One can use this formula to show that no matter how fast Gertrude and Pablo move in opposite directions on the track, they will never appear to separate at the speed of light.

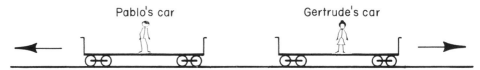

FIGURE 4.4

Box 4.2
COMBINING SPEEDS IN RELATIVITY THEORY

For the situation sketched in figure 4.4, Newton's laws of motion would state:

Speed of Pablo with respect to Gertrude =
Gertrude's speed with respect to you +
Pablo's speed with respect to you,

or

$$(3/4)\,c + (3/4)\,c = 1.5c,$$

where c is the speed of light. Or even more briefly,

$$V_{\text{PABLO SEEN BY GERTRUDE}} = V_{\text{PABLO}} + V_{\text{GERTRUDE}}$$

where V_{PABLO} is Pablo's speed with respect to you and V_{GERTRUDE} is Gertrude's speed with respect to you.

For the situation sketched in figure 4.4, Einstein's model says (as a consequence of the Lorentz transformations):

$$V_{\text{PABLO SEEN BY GERTRUDE}} = \frac{V_{\text{PABLO}} + V_{\text{GERTRUDE}}}{1 + \dfrac{V_{\text{PABLO}}\,V_{\text{GERTRUDE}}}{c^2}}$$

and, substituting $V_{\text{GERTRUDE}} = (3/4)\,c$ AND $V_{\text{PABLO}} = (3/4)\,c$ we have,

$$V_{\text{PABLO SEEN BY GERTRUDE}} = \frac{(1.5)\,c}{1 + \dfrac{9}{16}}$$

$$= 0.96\,c$$

The "Twin Paradox"

Suppose we have identical twins (biological clocks) moving past each other at a uniform speed on the railroad track. Twin #1 looks at twin #2 and measures a slower aging process than he is undergoing himself. But twin #2 looks at twin #1 and determines that twin #1 is aging at a slower rate than he. So each twin says that the other is aging more slowly than himself. This is indeed predicted by special relativity, but it does not sound quite logical. How can each twin age less than the other?

Here is a usual statement of this "paradox." Let the twins move apart for some time so that the difference in aging measured by each is substantial; then bring the twins back together and let them stand side by side and compare their appearances. Each will say that his sibling is younger than he, but both cannot be right. Which twin is, in fact, older, or is there any age difference seen at all?[9]

The resolution of the "paradox" depends upon a careful analysis of how the experiment actually can be carried out. When the twins separate on the tracks it is true that each will claim the other is aging more slowly than he, but notice that this is a conclusion based on *measurements* made of the other twin. Unless we change the direction of motion of at least one twin (that is, accelerate the twin), the two twins will continue to separate forever, and we cannot *directly* compare their two ages (this is like bringing two events to the same point in space so that simultaneity can be determined directly). But since we do want to compare the twins' ages directly by comparing their simultaneous appearance, suppose that we accelerate twin #1 and bring him back to stand at rest with respect to twin #2. Notice that now the two twins no longer have identical histories. One twin has undergone an acceleration (and the twins can tell which has been accelerated because the accelerated twin moved in a noninertial frame for a time and so experienced some sort of force, as described in section 2.4). As we will see in the following chapter, in accelerating, twin #1's clock will be slowed relative to twin #2. When the twins are standing at rest with respect to one

---

[9] Actually, experiments very similar to this one have been performed using "twin" atomic clocks; a technical discussion may be found in two papers by J. C. Hafele and Richard E. Keating, "Around-the-World Atomic Clocks: Predicted Relativistic Time Gains," *Science* 177 (14 July 1972): 166–167, and "Around-the-World Atomic Clocks: Observed Relativistic Time Gains," *Science* 177 (14 July 1972): 168–170. The same experiment is given a more thorough (though still technical) discussion in J. C. Hafele, "Relativistic Time for Terrestrial Circumnavigations," *American Journal of Physics* 40 (January 1972): 81–85. The results are consistent with relativity theory.

another, the twin who suffered the acceleration will have aged less than the other.[10]

The key to the "paradox" is that the histories of the twins moving relative to one another must differ if we are to compare their ages directly by standing them side by side after they have been moving for a time. The key historical fact is that one twin has gone through a noninertial episode, and that will cause a permanent difference in clock readings. But if, for example, in bringing the twins back to stand side by side both twins had been accelerated in exactly the same way, then their histories would be identical and there would be no difference in their ages.

The "Paradox" of Length Contraction

Recall our discussion of your measurement of Gertrude's length as she sped by you on the track (see section 3.8 and figures 3.19–3.23). We showed that Gertrude's length will be measured by you to be contracted, and you will conclude that Gertrude's car just fits into the tunnel on the track. But now consider things from Gertrude's point of view. She observes the tunnel moving past her at a uniform speed, and her measurements lead her to conclude that the tunnel is contracted and is too short to contain her car.

For the sake of argument, let's substantially increase Gertrude's speed with respect to the tracks, so that you now measure Gertrude's length to be half the length of the tunnel. On the other hand, Gertrude measures the tunnel to be much too small to contain her car. As in the "twin paradox," you each make measurements that seem mutually contradictory. Which of you is correct? Will Gertrude's car fit or will it not fit in the tunnel?

An imaginary physical confrontation of the two points of view may be arranged in the following way (figure 4.5). You block the exit of the tunnel with a wall of reinforced concrete, and you arrange a steel portcullis to fall and cover the tunnel entrance as soon as the rear of Gertrude's car passes into the tunnel. This experiment could be run in principle (although it would certainly be Gertrude's last train ride). You say Gertrude's length will easily fit into the tunnel so that the falling portcullis will not touch her car; she says that the tunnel is much too short to hold her car so that the

---

[10] We have invoked a result of general relativity that is discussed in the following chapter to explain this "paradox." Actually, the same result can be argued using only special relativity theory, but the details are more complex than our treatment warrants. See French, *Special Relativity*, pp. 154–159, for a discussion aimed at undergraduate physics students.

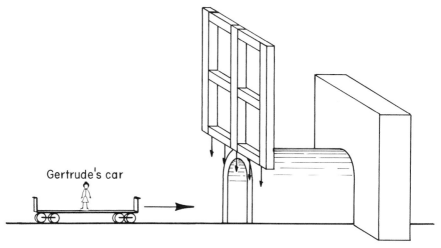

FIGURE 4.5

portcullis will hit her train: the question is, What would "really" happen when the portcullis falls?

In a sense both you and Gertrude are correct in your conclusions, although the portcullis will, in fact, close to confine Gertrude's entire car (or what is left of it) in the tunnel. Let us explain the resolution of the "paradox."

Refer to your original length measurement discussed in section 3.8. If Gertrude's car were brought back to the location of the tunnel and placed at rest on the tracks beside it, you (and Gertrude) would see that her car is longer than the tunnel. Because of the length contraction predicted by special relativity, you measured the car's length to be equal to that of the tunnel as it moved past you at speed v. From Gertrude's perspective, at speed v the tunnel, already shorter than her car when at rest beside her, becomes shorter still as she moves past it so that there is no possibility for her car to fit inside with the portcullis closed.

Now, when you determine that the rear of Gertrude's car enters the tunnel, the portcullis closes and the car is trapped in the tunnel. This *is* the result of the experiment, and we easily understand it from your perspective. Gertrude cannot disagree with this result (either it happens or it does not when the experiment is performed), but how does she reconcile it with her measurements?[11] Gertrude enters what she has measured to be a very short

[11] Here we follow the excellent discussion provided by Wolfgang Rindler, *Essential Rel-*

tunnel; she soon strikes the closed end of the tunnel, which stops the motion of the front of the car. But the rear of her car will keep right on going for a while; it will "learn" of the collision at the front end only after a finite amount of time (after all, communication of any kind is possible from the front to the rear of the car at a speed at most equal to that of light). Meanwhile, Gertrude's car is buckling and splintering from the front end toward the back due to the force of the impact with the reinforced concrete wall. Eventually the rear of Gertrude's car "gets the message" that the front has been stopped by the tunnel block, but not before the car has been buckled up inside the tunnel and the portcullis is closed.

We have made the highly probable assumption that Gertrude's car would be destroyed by the impact with the concrete wall. The actual condition of Gertrude's car, however, is of no importance to the discussion. Even if the material of Gertrude's car were strong enough to withstand the force of the impact, her car would still deform so as to be trapped inside the closed tunnel. It might bend out of shape like a bow, or become compressed like a spring squeezed from both ends. If the portcullis were to open after the experiment, her car would "spring back" to its normal length if it were not damaged.

So we say that both you and Gertrude are correct. The car is too long to fit in the tunnel, at least without being deformed (Gertrude's conclusion), and yet the car does fit into the tunnel with the portcullis closed (your conclusion).

4.7 WHEN IS "NOW"?

By this time it should be clear that in relativity theory, one must be exceedingly careful to specify the frame of reference from which measurements of the times and locations of events are made. One must also be clear about the meaning of terms related to measurements (recall Einstein's careful definitions of the time for an event). For example, consider the situation shown in figure 4.6. You are again standing beside the railroad track along which we have located a number of flashbulbs. At any time (as read on your clock) you can ask, "What happens now?" In asking this question, you can

---

ativity (New York: Springer-Verlag, 1977), pp. 41–42, who uses the example of a man trying to fit a twenty-foot pole into a ten-foot garage by running with the pole into the garage at 86.6 percent of the speed of light so that the length of the pole is contracted to ten feet from the perspective of someone at rest in the garage. Rindler's book is intended for readers with a background in physics and calculus.

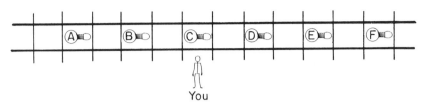

FIGURE 4.6

mean one of two things from the standpoint of relativity theory. There is "now" in the sense of simultaneous events on the common time system. In this sense, suppose that "now" (marked by a certain reading of your clock on the common time system) all of the flashbulbs fire simultaneously (where we are using Einstein's definition of the word "simultaneously"). We can answer your question "What happens 'now'?" by saying that at the instant "now" all the bulbs went off simultaneously in the common system of time in use along the track. Of course this is not what you actually *saw* when you said "now." Sometime after the bulbs went off the light from them began to reach you. You first saw bulb C fire, and then the light from bulbs B and D reached you at the same time, and still later you saw the light from bulbs A and E. So you did not receive information about what happened "now" until after it happened, and notice that the receipt of that information did not occur in one instant: it was spread out over an interval of time. In fact, if the railroad track were long enough and covered with flashbulbs, you could still be receiving signals of what happened "now" even if that "now" were decades in the past.

There is another sense in which you can use the word "now." In this sense you mean what you actually *see* at that instant of time. Right "now" you might see bulb C flash. A little later on, right "now" you would see bulbs B and D flash, and so on.

Of course we also know that an observer moving past you will see events differently than you. So not only must the sense of the word "now" be defined with care, but we must always be clear about the frame of reference to which the "now" refers.

4.8 THE RELATIVITY OF TEMPORAL ORDER AND CAUSALITY

Figure 4.7 shows two of the flashbulbs in figure 4.6; we will use this figure to illustrate another important point to bear in mind when dealing with

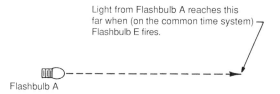

Light from Flashbulb A reaches this far when (on the common time system) Flashbulb E fires.

Flashbulb E is sufficiently far from Flashbulb A that the light ray from Flashbulb A does not reach the position of Flashbulb E by the time Flashbulb E fires.

Flashbulb A

Flashbulb E

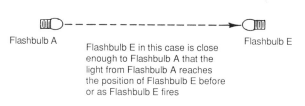

Flashbulb A

Flashbulb E in this case is close enough to Flashbulb A that the light from Flashbulb A reaches the position of Flashbulb E before or as Flashbulb E fires

Flashbulb E

FIGURE 4.7

relativity: the time order of events and the possibility of a causal connection between them are related to the distance separating the events. You determine that the flashbulbs fire simultaneously on your common system of time. Gertrude, moving by on her car, will make measurements that lead her to conclude that the bulbs do not fire simultaneously on her common time system.

In fact, according to the special theory of relativity, if the bulbs are far enough apart along the track (and we will soon say just how far is "far enough"), Gertrude's measurements could lead her to conclude that bulb A fired before bulb E, simultaneously with bulb E, or after bulb E, depending upon how Gertrude moves along the track with respect to you (both her speed and her direction of travel are important). So the measured temporal order of some events (those far enough apart in space) depends on the state of motion of the observer. We state this result of relativity theory without proof here because a proof requires a rather detailed analysis of the Lorentz transformations. In appendix A we use Minkowski diagrams (but no algebra) to prove this result. For the present discussion it is important to bear in mind that we are not speaking here of what order of events one actually *sees* at any instant (what one actually sees will be treated in the next section of this chapter). We are here addressing the question of what one *determines from measurements* to be the temporal order of events on the common time system. Measurements by Gertrude (one inertial observer) can lead her to conclude that the firing of flashbulb A (event A) precedes the firing of flash-bulb E (event E); Alice (a second inertial observer moving differently from Gertrude) can conclude that the same event A follows event E. You (a third inertial observer at rest on the track) conclude that A and E fire simultaneously. The "true" temporal order of events becomes an ambiguous concept for events that are sufficiently separated in space.

But it is not true that the measured temporal order of all events can be reversed by moving in different ways. We have been careful to say that only observations of events sufficiently separated in space can lead to contradictory conclusions regarding their temporal sequence. Let us now be specific about how far apart these "sufficiently separated" events must be.

In figure 4.7 we show beams of light sent out from flashbulb A at the instant that A fires. When the first event E occurs, the light beam from event A has not had enough time to reach the position of E. In this case flashbulbs A and E are "sufficiently separated" that different inertial observers may reach different conclusions about the temporal order of events A and E. On the other hand, the second event E is close enough to event A that the light ray from event A reaches the position of event E by the time event E happens. For such events all observers will agree on the temporal sequence of events A and E: A will precede E.

We have stated the results of the last two paragraphs without proof, but we wish to point out that they are necessary results if our usual sense of causal connections is to hold. Suppose there is some causal connection between the two events A and E shown in figure 4.7. If event A is to cause event E, then some sort of signal must travel from event A to the location of event E in order to trigger event E. As we have already remarked in this chapter, no known signal can travel faster than the speed of electromagnetic radiation (light). If events A and E are sufficiently close in space that a light signal has time to travel from the location of one event to the location of another to trigger it (as in the second example above), there can be a causally imposed temporal order on the two events. But, as we have remarked, the Lorentz transformations can be used to prove that such proximate events will only be measured to occur in the correct causal order no matter what the motion of the observer. In other words no inertial observer can move so that effects are measured to occur before their causes.

If, however, events A and E are so far apart in space that light (or any other signal) does not have time to travel from one event to the other during the time interval between their occurrence (as in the first example in figure

4.7), the events cannot be causally related because there is no way for them to "communicate" in time for a causal connection. As we already remarked, the Lorentz transformations demonstrate that such events may be measured to occur in any temporal sequence, depending upon the direction and speed of the inertial observer's motion.

Special relativity predicts disagreements among inertial observers about where and when things occur, but there can never be a disagreement about whether or not a causal relation is possible.[12]

4.9 How Objects Appear When Moving By at Very High Speeds

We have noted that as objects move by an observer at very high speeds, their measured lengths are contracted in the direction of travel (it turns out that no such length contractions take place in directions other than that of the motion of the object). How would such "contracted" objects appear if it were possible to look at them or to photograph them?

It is interesting that this question received no thorough analysis until 1959.[13] Nor is the question an easy one to answer, for a number of things happen at once to introduce several kinds of "distortions" in what one actually *sees*. Here we will indicate the general issues to be considered in arriving at an answer, and then we will provide some specific examples.

First some general remarks. In our discussion of special relativity, we have used the verb "to measure" in describing determinations of lengths and times made by various "observers." At the same time we have avoided the verbs "to see" and "to observe" in this connection unless the measurements did involve the actual receipt of a light signal. The verb "to measure" refers to what one determines as length and time values using an array of observers and clocks along the track synchronized to a common system of time. These *measurements* do not refer to what one would actually see

---

[12] In appendix C we discuss possible consequences of the existence of *tachyons*, hypothetical particles that move faster than light. Such particles could permit causally related events to occur in either order. This fact has been used as an argument against the existence of tachyons.

[13] See French, *Special Relativity*, pp. 149–152; the footnote on page 150 of French's book gives references to original work. Our discussion in this section owes much to French's book, to Rindler, *Essential Relativity*, pp. 57–60, and to the paper by G. D. Scott and M. R. Viner, "The Geometrical Appearance of Large Objects Moving at Relativistic Speeds," *American Journal of Physics* 33 (1965): 534–536, and references 1–7 therein. All of these treatments are intended for advanced students of physics.

with the eye or photograph with a camera.[14] We mentioned this fact early in our discussion of relativity theory, although we did so only in passing, and we have avoided discussion of what one would really see in high-speed motion to reduce confusion in our discussion of relativity itself. Let us begin our discussion of what one actually sees in relativity experiments by analyzing a familiar situation, as we have done several times before.

We return to our place at the side of the railroad track along which Gertrude moves at some uniform speed v. Consider two events in figure 4.8: event 1 is Gertrude's passage by your position along the tracks; event 2 is Gertrude's passage by the position of the railroad crossing sign at the side

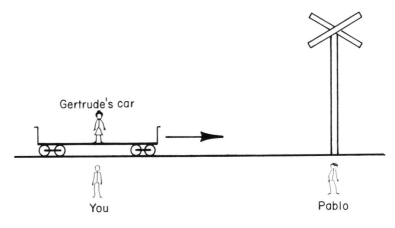

Gertrude's car

You   Pablo

FIGURE 4.8

[14] For example, in the discussion of Gertrude's moving light clock in section 3.7 we use an array of "observers stationed at rest along the side of the track" to "report to you the time required for the pulse of light in Gertrude's clock to make a round trip." Later on (section 3.8) we discuss the determination of Gertrude's length with respect to that of the tunnel by you and Gertrude. In interpreting figures 3.19 to 3.23 we have replaced an array of observers with mirrors that permit you and Gertrude to receive light signals from the crucial events. Because both sets of mirror apparatus are constructed so that the distance from the mirrors to your and Gertrude's eyes is the same, the simultaneity (or the lack of simultaneity) of the two events can be determined by "seeing" if the flashes of light from the two events reach your eye simultaneously. This apparatus was used in section 3.8 because although it yields conclusions fully equivalent to the results of making a measurement with an array of observers, it is far easier to visualize and to interpret. On the other hand, the times at which you and Gertrude *see* the flashes are not the common times at which the corresponding events occur. The apparatus permits a direct determination of simultaneity but not of common time—for that you need an array of observers.

of the tracks. Recall again what we mean by the times of these events. By Einstein's definition, the time of an event is the time given by a clock synchronized (according to his stated procedure) to the common system of time for all who share your state of rest with respect to the tracks, and located at the same point in space at which the event occurs. In figure 4.8, the time of event 2 (Gertrude's passing the crossing sign) cannot be observed directly by you because you are not at the same location as event 2; instead you must rely on Pablo, your helper, who is at the same position as event 2, and who determines the time of the event there with a synchronized clock. You learn of the time of event 2 by a report of some kind that you receive from Pablo; that is, your information must be second-hand—this is a matter of necessity.

We can represent these same two events on a Minkowski or spacetime diagram, as shown in figure 4.9. You, the sign, and Pablo stay fixed by the side of the tracks; the corresponding world lines are, accordingly, vertical, meaning that for all time values you have the same position. We also show the world line for the front of Gertrude's car as it moves along the track at the speed v (the solid line). Notice that Gertrude's position coincides with yours at the event we have called 1; similarly, the front of her car passes the crossing sign at the event labeled 2 in the diagram.

From this diagram we can tell the times of events 1 and 2 on the common time system. These we read as 2 and 3.4 time units, respectively. Again, these are the times for the events on the common time system as recorded and then reported by you and Pablo by the tracks.

Let us now ask what you *see*. First you see event 1, and because you happen to be right at the position of this event, the time of this event is the same as the time that you read on your clock when you actually see the event. But not so for event 2. Look at the spacetime diagram again. Event 2 occurs some distance from you. Therefore it takes light some amount of time to travel the distance from the sign to your location, and so you will see event 2 at a time on your clock later than that reported by Pablo as the time of the event. We can use the spacetime diagram to determine just when you will see event 2. At event 2, draw a dashed line representing the motion of the flash of light showing that the front of Gertrude's train has just passed the sign. Here and in what follows the world line for any flash of light is a dashed line inclined at 45 degrees to the time or distance scales.[15] The flash

---

[15] As discussed in detail in appendix A.2, throughout this book we have arbitrarily chosen the units of time and distance measurements in our Minkowski diagrams so that world lines of light are inclined 45 degrees to the time and distance scales.

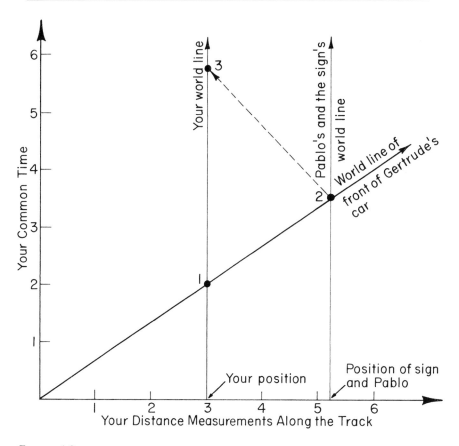

of light marking event 2 is seen to cross your position by the side of the tracks at event 3. Event 3 marks your first sight of event 2.

We have been rather long-winded about all of this because it is so important to understanding what observers *see* of natural phenomena, as opposed to what they will *measure*. Relativity theory (through the Lorentz transformations) refers to measurements with clocks and rulers of common times and positions of events. By considering how light leaves these events and travels to our eyes (or to the lens of our camera) we can also use relativity theory to deduce what you and any other observer will actually see. This is just what we have done in the Minkowski diagram in figure 4.9; we have added the appropriate light-ray path to determine when and where you will see event 2. This act of viewing event 2 we call event 3. By following a similar procedure (using Minkowski diagrams or the Lorentz transformation equations) for various sorts of events one can reach general conclusions concerning the "distortions" that are *seen* in objects moving by at high speed.

In general there are three effects at work that serve to "distort" what one actually sees of a moving object. Let us consider each of these in turn, and to make the discussion specific we will again use Gertrude moving by you in her car at a substantial fraction of the speed of light (to make the effects more noticeable). Figure 4.10 is the sort of bird's-eye diagram of Gertrude and the track that we used earlier in discussing your measurement of Gertrude's length (figures 3.19 to 3.23). We will suppose that you are standing

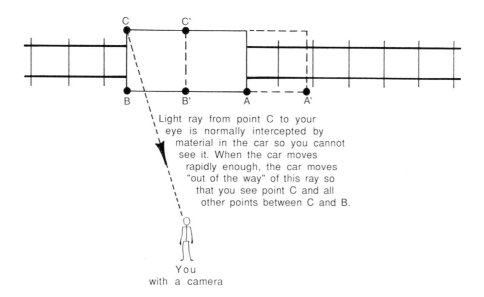

Light ray from point C to your eye is normally intercepted by material in the car so you cannot see it. When the car moves rapidly enough, the car moves "out of the way" of this ray so that you see point C and all other points between C and B.

You
with a camera

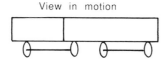
View in motion

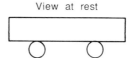
View at rest

FIGURE 4.10

beside the track with a camera having an extremely fast shutter so that you are able to "stop" the image on the film without appreciable blurring. When the center of Gertrude's train appears in your viewfinder to be directly opposite your position along the trackside, we suppose that you trip the shutter and take a picture. The discussion to follow will refer to the photograph that results from this procedure.

Effect #1: The Lorentz Contraction

First remember from our discussion in section 3.8 that your measurements will determine the length of Gertrude's car to be less than she knows it to be; in fact, you will also *see* Gertrude's length contracted relative to what she claims her length to be. Imagine the tunnel shown in figure 3.19 to be located over the track with its center directly opposite your camera lens; your camera now replaces the mirror apparatus of section 3.8 that permitted you to view both ends of the tunnel without moving your head. As in section 3.8, your picture shows both ends of Gertrude's car simultaneously fitting into the ends of the tunnel. In other words, your picture shows that the length of Gertrude's car just fits into the tunnel. The side of her car therefore appears shorter to your camera when the car is in motion past you than it would be if the car were photographed at rest beside the tunnel.[16] But that is not all.

Effect #2: An Apparent Rotation

Look carefully at the top portion of figure 4.10 to see what the light rays are doing as they leave the car and head toward your camera. Consider in particular light rays from point C at the rear corner of the car. We have drawn one ray (as a dashed line) emitted in your direction; but the side of the car (with corners A and B) stands between you and the path of this ray, so you normally cannot see light from point C or from any other point along the rear edge of the car between points B and C. The car itself blocks your view of these points.

But now let Gertrude begin moving past you at a very high speed. It takes light from point C a certain amount of time to travel across the width of the

[16] Some of the original work done on the appearance of objects moving past an observer at high speed seemed to suggest—incorrectly—that the Lorentz contraction would not be observable. By 1961, however, it was realized that the contraction would be visible. Readers should be cautious in consulting pre-1961 literature in this regard.

car. When the car was at rest, the matter of the car interrupted the light from point C and prevented you from seeing it, but now the rear edge of the car moves out of the way rapidly enough for the ray to clear the back of the car and move to your eye. By the time the light ray from point C has moved across the width of the car, the end of the car that used to block that ray when the car was at rest has moved on (for example, point B has moved on to point B'), well out of the way of the ray's path. In other words, you can see the far corner of Gertrude's car marked by point C. Similarly, you are able to see light from all of the other points along the rear edge of her car from B to C. The net result is that while you see the side of her car contracted, you are also able to see a foreshortened view of the rear edge of the car. This is sketched in the lower portion of figure 4.10. In effect, you no longer see Gertrude's car ''side-on'' when she moves past you. The car effectively appears somewhat rotated and now presents a kind of ''cubist'' image of itself. In fact, it turns out that if you view the car from a great enough distance, the side and the rear edge appear foreshortened in exactly the way that you would observe if the car were simply rotated a bit. The faster the car moves along the track, the greater the amount of the observed ''rotation.''

Effect #3: The Travel Time for Light

Suppose that the object moving past you now is a long, rigid bar standing vertically on Gertrude's flatcar. In the following discussion remember that the length contraction is still operating, so that the bar appears skinnier than it would at rest with respect to you; in addition ''rotation-type'' effects discussed above are present. Figure 4.11a is a view of your camera and the bar looking *along the track* from the ground level. We have located three points on the bar and have drawn dashed lines to indicate the path that light will take in moving from the bar to your camera. Light from point B at the middle of the bar will reach your camera soonest since it has the least distance to cover. Light from A and C will require more time.

You take your snapshot when you see point B pass your position along the track. Of course the light that you see (and that strikes your camera lens) actually left point B on the bar some instants before. By the same token the light simultaneously striking your lens from points A and C had to have left the bar before the light from point B, since these points are more distant from the lens than point B. This means that the light you photograph from points A and C left the bar when it was even farther away from you on the

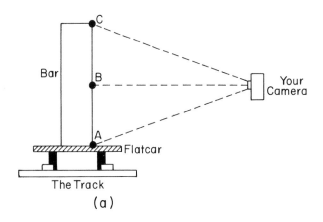

The Track

(a)

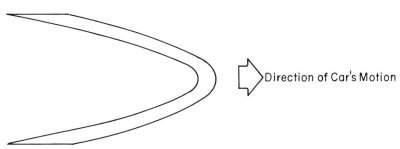

This is how the bar appears in the photograph made by the camera.

(b)

FIGURE 4.11

track than when the light from point B left the bar. So you see the center of the bar (point B) directly across from you while points to either side of B (such as A and C) appear to have some distance to travel before being opposite you. The amount of the "distance to travel" increases as you view points farther along the bar from point B. Thus the bar will appear bowed, with the center (point B) sticking out in a forward direction along the tracks and points to either side of B appearing farther behind along the track (it can be shown that the shape of the "bow" is that of a hyperbola). A photograph of the bar taken with the camera would look like the sketch shown in figure 4.11b.

So even though there is no length contraction perpendicular to the direction of motion of the bar, because of the travel time for light the vertical bar appears bent or "swept back" due to its motion.

The same effect applies to the appearance of Gertrude's car. Let's take another bird's-eye view of the car as it passes directly in front of you (figure 4.12a). Pablo has painted some evenly spaced spots on the side of the car facing you and your camera to make the "distortion" easier to see. Five markers lettered A', A, B, C, and C' have been fixed to the trackside to aid our discussion.

To understand the appearance of the photograph you make of Gertrude's car, it is necessary to describe carefully the sequence of events leading to the opening of the camera's shutter. The shutter will open momentarily when light reaches the lens from the coincidence of the center of the car with marker B (the marker on the track directly opposite the camera lens). This defines the time at which the shutter opens, and it is important to bear in mind that only light striking the lens in the moment that the shutter is open will be recorded on the photograph. Of course the light from the cen-

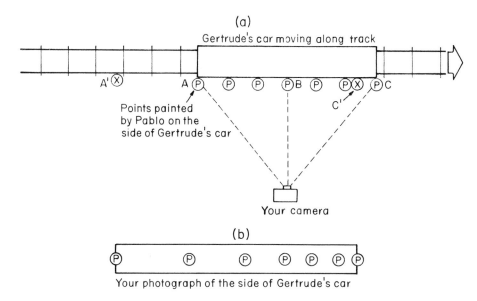

(a)

Gertrude's car moving along track

Points painted by Pablo on the side of Gertrude's car

Your camera

(b)

Your photograph of the side of Gertrude's car

FIGURE 4.12

ter of the car at point B requires some amount of time to reach the lens; when the car's center is at marker B on the track, the front and rear of the car are at markers C and A, respectively, and light from these two markers has to travel farther to reach the lens than does the light from marker B. Therefore, light from the front and rear of the car at markers C and A will arrive at the lens later than the light from the center of the car at B and after the shutter has closed. So although the photograph will show the car's center at marker B, and although when the center of the car is at B the rear and front of the car are at A and C, the car's two ends will not be photographed at markers C and A. The light from these two points simply does not reach the lens in time to pass the open shutter.

Of course the image of the car will have a front and rear, but they will not appear along the track at markers A and C where we expect them. To reach the lens at the same time as light from the car's center at marker B, light would have to leave the front of the car before the front passed marker C—at a marker such as C'. Light from the rear of the car, to reach the lens with the light from the car's center at marker B (so as to pass the open shutter), must be emitted at a marker such as A'. So the light emitted when the rear and front of the car are at markers A' and C' define the rear and front of the photographed image.[17]

Figure 4.12b shows the resulting snapshot of Gertrude's car. The center of the car is at the center of the picture (at marker B, not shown in the photograph). The pattern of Pablo's spots (drawn to be evenly spaced on the side of the car) appears "compressed" in front and "stretched out" in the back.

In figure 4.13 we have combined all three of these effects to illustrate how several objects would photograph as they pass by a camera at high speed. Imagine the objects to be mounted on Gertrude's car and that you are standing with a camera beside the track as in figure 4.12a. The camera shutter is tripped when the center of each object passes directly in front of

---

[17] Notice that marker A' is located farther from marker B than is marker C'—that is, light must leave the rear of the car at an earlier time to pass the shutter than does light leaving the front of the car. The reason for this can be understood by thinking carefully about figure 4.12. We have already argued that to pass the camera shutter, light from the front of the car must leave when the front is at marker C'. When the front of the car is at C' the rear is just slightly to the left of marker A, but notice that light from the rear at this marker has farther to travel to the lens than does light from the front at marker C'; therefore, light from the rear of the car must leave a marker even farther back along the track to reach the lens with the light from the front of the car at C'. Point A' is such a point.

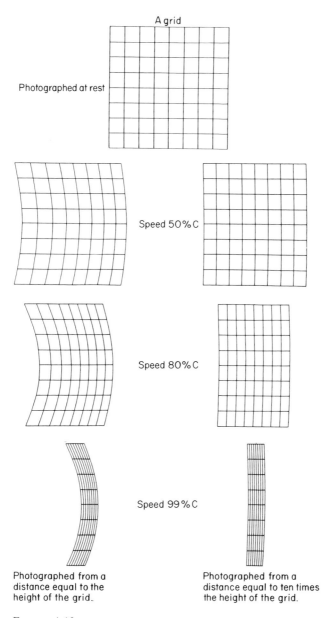

FIGURE 4.13a.

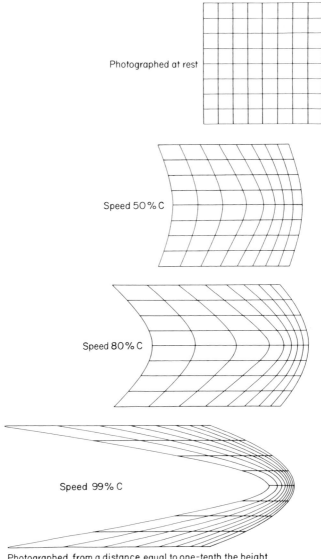

Photographed from a distance equal to one-tenth the height of the grid.

FIGURE 4.13b.

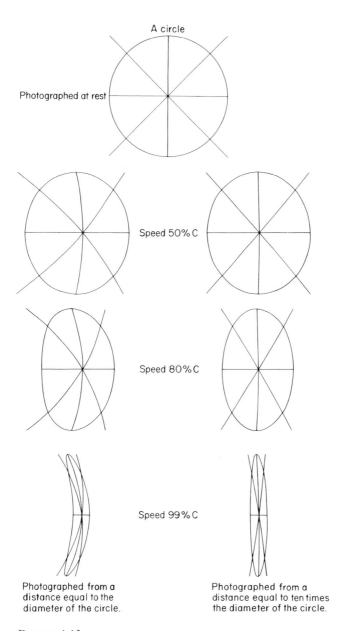

A circle

Photographed at rest

Speed 50% C

Speed 80% C

Speed 99% C

Photographed from a
distance equal to the
diameter of the circle.

Photographed from a
distance equal to ten times
the diameter of the circle.

FIGURE 4.13c.

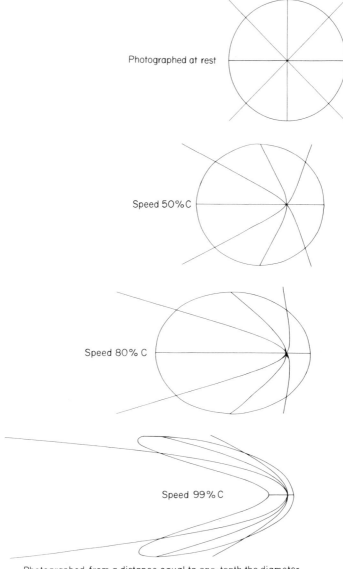

Photographed at rest

Speed 50% C

Speed 80% C

Speed 99% C

Photographed from a distance equal to one-tenth the diameter of the circle.

FIGURE 4.13d.

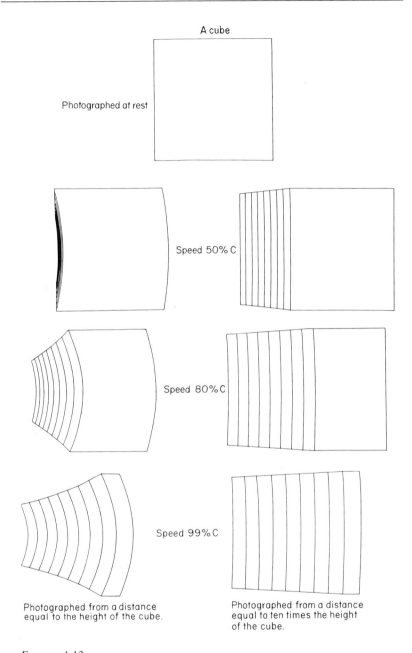

Photographed from a distance
equal to the height of the cube.

Photographed from a distance
equal to ten times the height
of the cube.

FIGURE 4.13e.

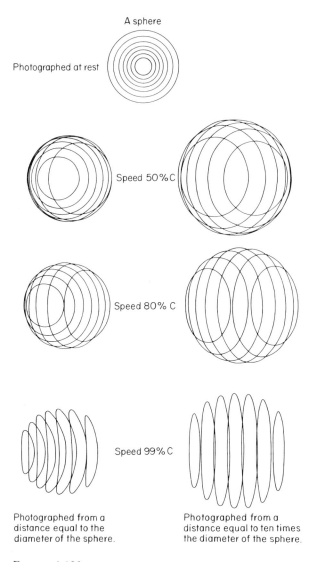

A sphere

Photographed at rest

Speed 50% C

Speed 80% C

Speed 99% C

Photographed from a
distance equal to the
diameter of the sphere.

Photographed from a
distance equal to ten times
the diameter of the sphere.

FIGURE 4.13f.

the lens. Despite the high speed of Gertrude's car, we assume the camera
shutter to be so fast that the image is not at all blurred. Photographs are
taken of two flat objects (a grid and a wheel with spokes) and three solid
objects (a cube, a sphere, and a cylinder).

The appearance of an object in the photograph depends on the speed at

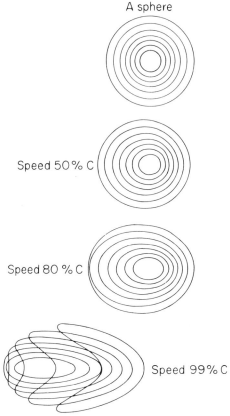

A sphere

Speed 50% C

Speed 80 % C

Speed 99% C

Photographed from a distance
equal to one-tenth the diameter
of the sphere.

FIGURE 4.13g.

which it passes the camera. To illustrate this effect, the objects are photographed passing the camera at each of three speeds: 99 percent, 80 percent, and 50 percent the speed of light. A photograph of each object at rest is also shown for reference.

The photographed shape of the objects also depends on the distance separating the camera and the object, so we show photographs taken from several distances: extreme close-ups (the distance from the camera to the object is only one-tenth of the height of the object—a photograph that would require an extreme wide-angle lens to capture the image of the entire object), moderate close-ups (from a camera distance equal to the height of the

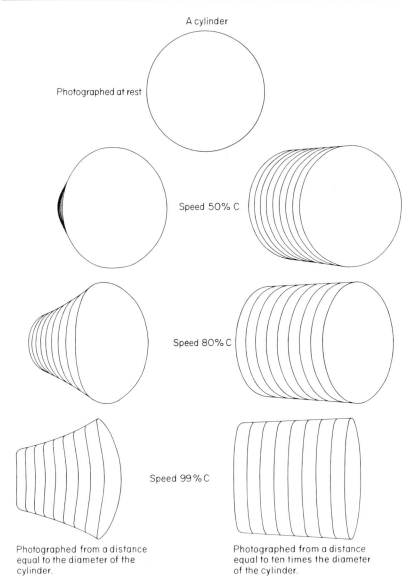

A cylinder

Photographed at rest

Speed 50% C

Speed 80% C

Speed 99% C

Photographed from a distance
equal to the diameter of the
cylinder.

Photographed from a distance
equal to ten times the diameter
of the cylinder.

FIGURE 4.13h.

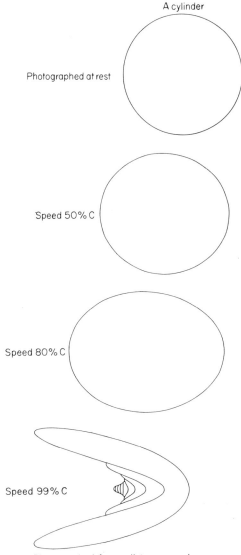

A cylinder

Photographed at rest

Speed 50% C

Speed 80% C

Speed 99% C

Photographed from a distance equal
to one-tenth the diameter of the cylinder.

FIGURE 4.13i.

object), and "distant" shots (with the camera separated from the object by ten times the object's height).

In the images of the grid and the circle, figure 4.13 clearly shows the compression and extension effect due to the travel time for light.

To clarify the "rotation" effects in the images of the solid objects, Pablo has painted a pattern of stripes on each. The cylinder is oriented "end-on" so that as it passes directly opposite the camera its circular faces are perpendicular to the line of sight. Seven stripes are painted at regularly spaced intervals around the circumference of the cylinder. When the cylinder is at rest, all the camera can see is one of the circular ends. As the cylinder moves, the relativistic "rotation" effect makes the pattern of stripes visible. The cube is similarly striped; it moves past the camera so that two of its faces are perpendicular to the line of sight. Seven equally spaced stripes have been painted on each of the other four faces. As with the cylinder, these stripes only become visible on the photographs when the "rotation" effect brings them into view. Unlike the cube and the cylinder, our sphere is transparent. If it were a transparent globe of the earth, the north and south polar points would fall exactly along the line of sight as the sphere passed directly opposite the camera; the pattern of seven stripes would then be circles of constant latitude, one of which falls on the equator. The points corresponding to the two poles are also shown in the figure.

In creating these "photographs" with a computer, portions of the solid objects more distant from the camera lens have been reduced in size to simulate the linear perspective a camera would impose.

Our illustrations show the expected result that "distortions" become more pronounced as the speed of the object passing the camera becomes larger. We also see that effect #3 discussed above (due to the travel time for light) is reduced at greater camera distances. This can be understood by looking again at figure 4.11. The farther the camera from the track, the smaller the difference in distance from the camera lens to points A, B, and C and so the smaller the "distortion" due to the travel time for light.

WE HAVE just illustrated some selected consequences of relativity theory. We realize that to some of our readers these consequences will be the most interesting part of this book and that they will wish that we had been more encyclopedic in our treatment. By way of excuse we can only say that we are not finished yet, for we still have to discuss the general theory of relativity wherein we will find "distortions" of an even more extreme and fundamental order.

5

THE GENERAL THEORY OF RELATIVITY

This suggestion of a finite but unbounded space is one of the greatest ideas about the nature of the world which has ever been conceived.

—Max Born

5.1 INTRODUCTION

We now turn to a discussion of Einstein's crowning achievement, the general theory of relativity. Our approach will be similar to that used in presenting special relativity: we will use the first few pages of one of Einstein's own papers to guide our discussion. But a bit of explanation is in order concerning this strategy. Einstein's special theory of 1905 burst upon the world in a single publication, complete and self-contained. We used the introductory pages of that paper to understand what Einstein did and what inconsistencies in classical physics had motivated his work. The general theory of relativity, on the other hand, was not announced suddenly in one publication. It appeared in November 1915 in *The Proceedings of the Prussian Academy of Sciences*,[1] but the problems involved had been discussed in a number of papers by Einstein beginning in 1907.

For our presentation of general relativity, we have chosen to be guided by the first seven pages of a paper Einstein published in March 1916, a paper written as a summary of the previous eight years' work on the theory and appearing in the same physics journal that had published his original relativity theory.[2] It is indeed fortunate for studies in the history of science

[1] A. Einstein, "Die Feldgleichungen der Gravitation," *Sitzungsberichte, Preussische Akademie der Wissenschaften* (1915), part 2: 844–847; this paper was communicated to the Academy in its session of November 25, although the paper was not published until December 2.

[2] A. Einstein, "Die Grundlage der allgemeinen Relativitätstheorie," *Annalen der Physik*

that Einstein published as he went along in his development of general relativity, because that series of papers from 1907 to 1915 represents an open record of the evolution of his thinking, including false starts, errors, and brilliant flashes of insight. We have no such record of his progress in formulating the original theory of 1905, and, as a result, questions concerning how and why Einstein did what he did then are more difficult to resolve. But our discussion is not a history; our concern will be with presenting the conceptual content and scientific contribution of the general relativity theory.

One more introductory remark is in order concerning the difference in our treatments of general relativity in this chapter and of special relativity in chapter 3. As we remarked in chapter 3, special relativity is based on rather simple and straightforward (if counterintuitive) ideas. Like Einstein in the first few pages of his original paper of 1905, we were able to present these ideas and develop them without depending on mathematics beyond arithmetic. Indeed, the sketch shown in figure 3.15 is only a short algebraic step from the Lorentz transformation equation for time. Similarly, we could extend our analysis of your measurement of Gertrude's length in figures 3.19 to 3.23 to obtain the other Lorentz transformation. In other words, our discussion was still quite close to the formal mathematical content of the special theory. The Lorentz transformations can be obtained with algebra, and this is all we avoided in chapter 3.[3] In our analysis we penetrated to a level just shy of casting the argument in mathematical terms and finishing the theory with formal mathematical (algebraic) manipulations.

This depth of presentation will not be possible in our treatment of general relativity. For one thing, the mathematics necessary to treat the general theory correctly goes far beyond algebra to the areas of tensor calculus and differential geometry, subjects normally studied by quite advanced students of physics. We will be able to talk *about* these subjects, but our discussion will not be as narrowly focused as in chapter 3. Another difficulty that we will discuss in detail in sections 5.6 and 5.7 has to do with the fact that general relativity denies the universal applicability of the comfortable, even "intuitive," Euclidean geometry that we have been taught to accept as "true" since childhood. Not only must we grapple with counterintuitive

---

49 (1916): 769–822. All quotations from this paper are taken from the translation to be found in *The Principle of Relativity*, pp. 109–164.

[3] Calculus is required for a full development of the theory; in particular, one needs calculus to prove that Maxwell's equations are invariant (retain their form) for all inertial observers.

relativistic effects similar to those discussed in chapter 3, but we must do so in a physical world that is geometrically strange. Casting that strangeness in ordinary language has proved to be difficult; even the mathematical treatment of these "non-Euclidean" geometries is difficult, and Einstein himself was forced to acquire with effort the necessary mathematical sophistication as he progressed in his development of the general theory.

5.2 THE NEED TO EXTEND THE PRINCIPLE OF RELATIVITY

Einstein titles his summary paper of 1916 "The Foundation of the General Theory of Relativity." It begins by referring back to the two postulates that lie at the foundation of his original theory of 1905. The first of these is the Principle of Relativity (discussed in section 3.6):

The laws of physics must be the same for all observers moving in inertial (non-accelerated) reference frames.

According to this postulate there are no privileged inertial reference frames. As we noted in chapter 3, this postulate seemed almost self-evident to Einstein; it was the way that nature *had* to work. But, at the same time, this principle holds for only a very special situation: unaccelerated motion (that is, for inertial observers, observers not undergoing a change in speed or direction of travel with respect to any given inertial reference frame). Einstein indicates the restrictive nature of the postulate by calling it in 1916 "the special principle of relativity"; similarly, he terms his relativity theory based on this postulate "the special theory of relativity" to indicate its limited domain of validity.

In his new, general relativity theory Einstein seeks to remove the restriction imposed by specifying unaccelerated motion; he does this by suggesting that there are no privileged reference frames of *any* sort and that, to use his words,

The laws of physics must be of such a nature that they apply to systems of reference in any kind of motion.[4]

Let us review briefly a situation we discussed in section 3.6 so that we can clearly distinguish what is being said here from what Einstein said in his special theory. You are in a railroad car that moves uniformly along the

[4] Einstein, *The Principle of Relativity*, p. 113.

track relative to the roadside (like Gertrude in her car). Suppose that the car and road provide a perfectly smooth ride. If you don't look out from a window of the car you will have no sense of motion and you will be unable to tell whether you are moving or at rest. Even if you were to look out of the window and see another train on an adjacent track moving past you, you could not tell (without looking out) whether you or the other train are at rest on the track. Similarly, Alice, sitting at a window of the other train, cannot tell whether her train or yours is moving over the tracks unless she looks out. This is just an application of the special principle of relativity.

But, as we pointed out in section 3.6, things would be fundamentally different for observers riding in accelerating systems of reference. Suppose, to continue our example, that the train in which you are riding forward suddenly speeds up. You will feel a force as a result of the acceleration; this force will push you back into your seat. Without even looking outside of your car, you can tell that you changed your state of motion and in which direction the change took place. If you measure the magnitude of the force that acts upon you, you can (using Newton's model of motion) determine the amount of the acceleration.

In other words, accelerated systems appear to be special or privileged: their accelerations seem to be "absolute" in the sense that in such systems an observer can determine her acceleration without reference to anything external (without looking out the window).[5] In unaccelerated (inertial) frames of reference it is impossible to determine an "absolute" state of motion or of rest. Einstein found this privilege of accelerated reference systems to be at odds with what he felt the world should be: "absolute" motion of any sort could not be admitted in physical theory. Thus, in the general theory, Einstein sought to extend the notion of equality of perspective established for nonaccelerating reference frames in special relativity by stating that *any and all* frames of reference are equally valid for expressing the laws of physics.

Of course Einstein now had a problem, the problem of the force you feel in your accelerating railroad car, a force that would seem to permit you to assert your state of acceleration in an "absolute" sense (without looking from your window). This problem Einstein resolved with what he called "the happiest thought of my life,"[6] a thought that developed into what is now called "the principle of equivalence."

[5] The word "absolute" in this context is not equivalent to Newton's use of "absolute" in reference to "absolute" space, time, and motion. See section 2.2. But see Einstein's use of the word as quoted later in this chapter.

[6] Pais, *Subtle Is the Lord*, pp. 178–179. Einstein's "happy thought" seems to have oc-

5.3 THE PRINCIPLE OF EQUIVALENCE

The principle of equivalence may be stated in the following terms:

In any small region of space, the effects produced by gravitation are the same as those produced by an acceleration.

We will explain why this principle is restricted to "small regions of space" presently, but first we illustrate the main content of the statement using another example (figure 5.1). Imagine yourself in a box in space. The box contains a complete laboratory for making measurements of your physical surroundings. You and your laboratory are so remote from any other object in the universe that you can neglect everything but yourself and the contents of the laboratory box, because everything else is too far away to influence you.

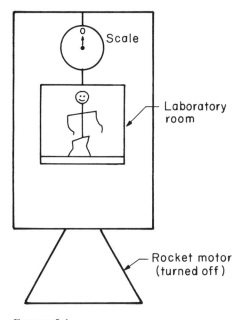

FIGURE 5.1

curred to him in 1907 at the very start of his effort to produce the general theory. Pais quotes Einstein: "I was sitting in a chair in the patent office at Bern when all of a sudden a thought occurred to me: 'If a person falls freely he will not feel his own weight.' I was startled. This simple thought made a deep impression on me. It impelled me toward a theory of gravitation."

There is a rocket engine attached to the bottom of your laboratory box but, to begin with, the engine is turned off and you are simply coasting through space with no acceleration.[7] Now the rocket engine turns on and applies a steady upward force to the laboratory (figure 5.2). In response to the force, the speed of the laboratory increases steadily. What do you experience as a result of the acceleration? You feel yourself pressed down to the floor of the laboratory. This is just the feeling of ''added weight'' experienced in an elevator when it begins to accelerate upward; it is also the very same sort of force that we discussed a few paragraphs above when you were riding in an accelerating train.

In thinking carefully about this situation, Einstein noticed that you would feel the very same sort of force in your laboratory if it were suspended above a planet and were undergoing no acceleration whatever (figure 5.3). In this case, the force would be due not to an upward acceleration of your car, but to the downward gravitational attraction of your mass by the mass of the planet. In fact, it turns out that if you did not have windows to look out of your laboratory to see whether or not you were suspended above a planet, you could not tell whether you feel the extra force because you are accelerating or because you are at rest and experiencing gravity.[8] In other words, the interpretation of the force you feel as due to an acceleration cannot be made with certainty. Like the unaccelerated traveler of special relativity, you can no longer make an assertion about your ''absolute'' state of motion.

This principle of equivalence also solves a longstanding puzzle of natural philosophy. Suppose we stand on a tall platform in your laboratory box suspended above the earth and drop a variety of objects to the floor. Any objects will do. We will choose Gertrude's railway car, a golf ball, and Alice. Drop all of these objects (initially in a state of rest with respect to the laboratory) from the top of the platform at the same time. Which of the objects will strike the floor first?

Since the time of Galileo it has been verified experimentally that (in a vacuum, so that only the force of gravity acts) falling objects move in an identical way; in particular, all of the objects dropped in the laboratory box

[7] Einstein used a similar picture in his popular book *Relativity*, written in 1916: ''. . . imagine a spacious chest resembling a room with an observer inside who is equipped with apparatus'' (p. 66). Einstein accelerated his laboratory ''chest'' by means of cables attached like elevator cables where we have used the more up-to-date rocket engine.

[8] Again, this statement holds only when your laboratory occupies a ''small'' region of space, as discussed below.

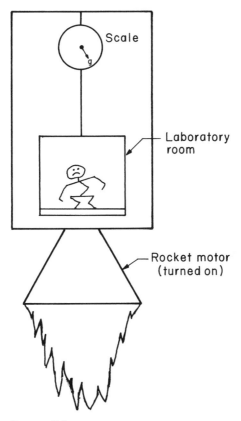

FIGURE 5.2

will hit the floor at the same time.[9] Why should all objects responding only to the force of gravity display identical motion?

Newtonian physics provided an explanation for this observed fact, although in doing so it presented an even deeper puzzle. The Newtonian explanation is this: One can use Newton's second law (discussed in section

---

[9] In what must have been the most expensive repetition of Galileo's seventeenth-century experiments, an Apollo astronaut stood on the surface of the moon (surrounded therefore by a vacuum) and dropped a rock hammer and a feather from the same height above the lunar surface. A camera recorded that the feather and the hammer fell in unison and hit the lunar soil at the same time. Some months later the experiment was repeated for a camera on earth using the same hammer and feather. The hammer hit the earth first—in fact, a gust of wind came by as the hammer and feather were released and blew the feather away before it could hit the ground.

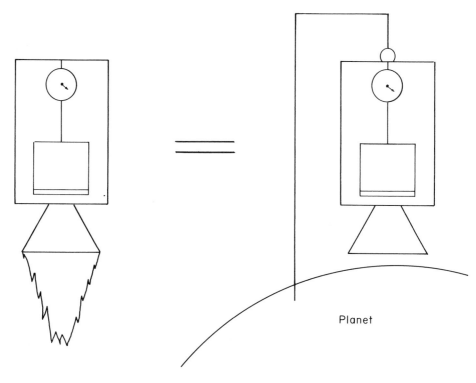

Planet

FIGURE 5.3

2.5) to predict the motion of an object having any inertial mass that is acted upon by some given force. But Newton also developed a model for the phenomenon of gravity that provided him with a way of predicting the gravitational force acting on any object. In Newtonian physics, gravity is an attractive force that acts between any and all pieces of matter in the universe. A simple formula, written as an equation, may be used to describe this force (box 5.1). The idea behind this Newtonian model is quite simple. It asserts that the degree to which objects attract one another because of gravity is measured by a property possessed by each piece of matter in the universe, a property called "gravitational mass." Gravitational mass can be measured for any sample of matter by using well-defined laboratory procedures. To determine the gravitational force that will act between any two objects according to Newton's model, multiply their measured gravitational mass values together and divide that number by the square of the dis-

Box 5.1
NEWTON'S MODEL OF GRAVITY

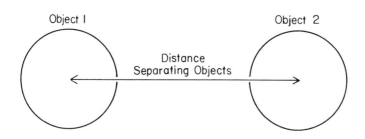

The force of gravity between two objects is proportional to the product of the gravitational masses of the two objects divided by the square of the distance separating them. In symbols:
    The force of gravity is proportional to

$$\frac{(\text{mass of object 1}) \, (\text{mass of object 2})}{(\text{Distance separating the objects})^2}$$

or even more briefly,

$$F = k \, \frac{M_1 M_2}{D^2},$$

where F is the force of gravity between the two objects,
        M_1 and M_2 are the gravitational masses of the objects,
        D is the distance separating the two objects, and
        k is a "constant of proportionality" chosen so that the desired units of force will result from the units chosen to measure mass and distance (see note 10 in section 5.3).

tance separating them. This bit of arithmetic will yield a value for the gravitational force acting between the two objects.[10] So the property of an object

[10] The number obtained in the calculation just described will be a value for the force that depends on the measurement standards employed for evaluating distance and mass. For example, one might measure distance in centimeters or feet and mass in grams or pounds. Obviously the force value calculated using Newton's formula will differ for different choices of these measurement standards. So that the calculated gravitational force will be given in terms

called its "gravitational mass" is the key to determining how that object responds to gravity. Now, despite the similarity of names, *gravitational* mass and *inertial* mass represent two very different properties of matter. Inertial mass measures the resistance an object offers to a change in its state of motion (see section 2.5). Gravitational mass measures the strength with which one object attracts another by the force of gravity. Newton determined (and subsequent experiments have verified) that the inertial mass value and the gravitational mass value for any given body are equal.[11] When this equality of mass values is incorporated in the algebraic combination of Newton's second law and his law of gravitation, the result is a simple prediction that all objects should move in exactly the same way when only the force of gravity acts. In other words, the explanation for the identical motion of all objects acted upon by gravity is contingent upon the equality of gravitational and inertial mass in Newtonian physics. But this only makes the puzzle more profound.

Why should two such disparate properties of an object as its gravitational mass and its inertial mass be described by the same number? This puzzle has bothered physicists since Newton's day. It certainly bothered Einstein. In his "Autobiographical Notes" he states:

> Now [in 1908] it came to me: The fact of the equality of inert and heavy [gravitational] mass, i.e., the fact of the independence of the gravitational acceleration on the nature of the falling substance, may be expressed as follows: In [the presence of] a gravitational . . . [force] (of small spatial extension) things behave as they do in a space free of gravitation, if one introduces in it, in place of an "inertial system," a reference system which is accelerated relative to an inertial system.[12]

Einstein's principle of equivalence at last solves the puzzle. In the laboratory box at rest above a planet, all objects released from the ceiling will fall to the floor in an identical manner due to gravity (the observed puzzle). Now let there be no planet nearby (no gravity) and suppose that the rocket

---

of an internationally accepted measurement standard of force, the number obtained in the force calculation is multiplied by what is called a "constant of proportionality," as shown in box 5.1.

[11] Recent experimental work is discussed in C. W. Misner, K. S. Thorne, and J. A. Wheeler, *Gravitation* (San Francisco: W. H. Freeman and Company, 1971), pp. 13–19. The treatment of the equality of gravitational and inertial masses in this reference assumes some knowledge of physics. The values of the two sorts of masses are found to differ by less than 0.0000000001 percent.

[12] Einstein in Schilpp, *Albert Einstein: Philosopher-Scientist*, I, 65.

motor turns on when all of the objects are released at the ceiling. The box will then accelerate past the free objects. An observer standing in the accelerating laboratory will see them all move to the floor in exactly the same way as under the force of gravity. The puzzle is solved.

Originally, Einstein thought of his principle of equivalence only in connection with the motion of objects: gravity and acceleration cannot be distinguished by observing the movement of objects from inside your (windowless) laboratory. But, in another of his brilliant flashes of insight, he extended the principle of equivalence to include not just the motion of objects, but all kinds of physical phenomena; that is, he suggested that all effects of acceleration—any and all effects—are fully equivalent to those produced by gravitation. This was a bold suggestion. It may or may not be an accurate description of nature. The only way to determine the domain of its validity (to use the terminology of chapter 1) is to test it against what is observed to be the case in experimental situations.

If functional, the principle of equivalence permits Einstein to assert the general validity of his principle of relativity. Forces that we used to think of as due to acceleration may now be ascribed to gravitation, or, to use Einstein's own words in an early (1911) paper on the developing general theory of relativity:

> This assumption of exact physical equivalence makes it impossible for us to speak of the absolute acceleration of the system of reference, just as the usual theory of relativity forbids us to talk of the absolute velocity of a system.

He goes on to note the added benefit of this principle of equivalence:

> It makes the equal falling of all bodies in a gravitational field seem a matter of course.[13]

And there is something else too. In the 1916 paper on the finished general theory, Einstein says,

> It will be seen from these reflections that in pursuing the general theory of relativity we shall be led to a theory of gravitation.[14]

[13] Einstein, *The Principle of Relativity*, p. 100.

[14] Einstein, *The Principle of Relativity*, p. 114. Actually Einstein realized very early in his development of a general relativity theory that Newtonian gravitation will not work within the context of his special relativity theory. The idea is this. As we have seen in section 4.5, according to special relativity inertial mass and energy are related according to the famous equation $E = MC^2$. Therefore, a more rapidly moving object (that is, a more energetic object)

In fact, Einstein's general theory of relativity *is* a theory of gravitation, one that supplants the seventeenth-century Newtonian model. Before going on to describe this Einsteinian model of gravity, however, we wish to make an epistemological point concerning the relation of Einstein's model to Newton's: in a very real sense Einstein's model of gravity does not "explain" the phenomenon of gravity any better than did Newton's model. Let us elaborate.

The Newtonian "law of gravity" is really a *descriptive* model of the phenomenon, and the description is quite good. However, even Newton was bothered by his model. He wanted (in the terminology of chapter 1) to comprehend a wider domain of validity than was accessible to his model; he wanted some further "explanation" of the phenomenon of gravity. Newton, along with many a Newtonian since, wondered how it is possible for masses to reach out across space and apparently exert forces on one another. In a famous letter to Richard Bently (1692–93, some five years after the first edition of the *Principia*), Newton writes:

That one body may act upon another at a distance through a vacuum, without the mediation of any thing else, by and through which their action and force may be conveyed from one to another, is to me so great an absurdity, that I believe no man, who has in philosophical matters a competent faculty of thinking, can ever fall into it.[15]

In the *Principia* itself Newton states:

We have explained the phenomena of the heavens and of our sea by the power of gravity, but have not yet assigned the cause of this power. . . . I have not been able to discover the cause of those properties of gravity from phenomena, and I frame no hypotheses; for whatever is not deduced from the phenomena is to be called an hypothesis; and hypotheses,

---

will have a greater inertial mass than when it is at rest. But as we have pointed out, experiments show that inertial and gravitational masses are equal, and the gravitational mass does not appear to depend on the motion of an object. So Einstein "abandoned as inadequate the attempt to treat the problem of gravitation . . . within the framework of the special theory of relativity." See Einstein, *Essays in Science*, pp. 79–80, and Schilpp, *Albert Einstein: Philosopher-Scientist*, I, 65. There were really two related issues of concern to Einstein at this time: the need to extend the principle of relativity to cover accelerated reference frames and the need to incorporate the phenomenon of gravity into his relativity theory. Pais, in *Subtle Is the Lord*, pp. 177–183, gives a technical account of the interplay of these factors in Einstein's development of general relativity.

[15] Quoted in Cajori's edition of the *Principia*, p. 636.

whether metaphysical or physical, whether of occult qualities or mechanical, have no place in experimental philosophy.[16]

Einstein did not provide an answer to this question raised by Newton concerning the "cause" of gravity, but he did create a much more accurate model of gravity than did Newton, and, perhaps more significant, Einstein's model is far richer in its consequences. It provides fundamentally new insight into the nature of the physical world. This added richness, like that of a truly great creative work in any field, makes the theory so attractive to specialists. Nevertheless, we will see after we have described Einstein's model for gravity that we are still left with a question of the sort "how is it possible . . . ?" just as with Newton's model.

There is yet another detail to clear up before we leave our discussion of Einstein's principle of equivalence and move on to our description of general relativity. Remember that the principle of equivalence begins with a qualification, "in any small region of space." Why is this qualification necessary, and just how small is "small?"

Consider the unusual situation sketched in figure 5.4. Here the rocket laboratory box is nearly as big as the planet above which it is suspended. Gravitational forces act between each of the two masses shown in the laboratory and the center of the planet. These forces are represented schematically by arrows in the figure. If the masses are released in the laboratory from some common height H above the floor, an observer in the laboratory will see them fall to the floor and, at the same time, move closer together. Were the laboratory to be accelerated upward with no gravity acting, the two masses would simply appear to fall in parallel lines right to the floor. Therefore, in this case, one could decide between an acceleration and the presence of gravity. The same experiment carried out in a much smaller room (or above a much larger planet) will still show one mass approaching the other to some extent; but if one chooses a small enough room, the approach of one mass toward the other (the departure of the masses from falling along strictly parallel lines) will be too small to be measured. Hence, we have the qualification in our statement of the principle of equivalence that it would hold in any "small region of space."[17]

---

[16] *Principia*, pp. 546–547.

[17] A physicist would say that the principle holds as long as the gravitational field involved is sufficiently "uniform," meaning that objects will fall along parallel lines; this uniformity may be achieved by considering smaller and smaller regions of space.

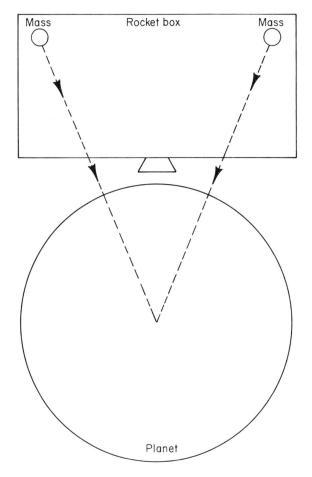

FIGURE 5.4

5.4 RESULTS OF THE GENERAL THEORY:
RELATIVITY IN THE MEASUREMENT OF TIME AND SPACE

Next in his summary paper Einstein points out how certain general conclusions of special relativity can be applied to situations in which observers
are accelerating with respect to one another. We have seen in chapter 3 how
time and distance measurements made by two observers in relative uniform
motion can differ. You at rest by the side of the track determine a size and
clock rate for Gertrude moving by you in her car. Gertrude measures different values for these. The differences arise because of the relative motion

of you and Gertrude. We stipulated in chapter 3 that you and Gertrude move *uniformly* with respect to one another, and that stipulation was invoked whenever we applied Einstein's second postulate (that the speed of light is the same for all uniformly moving—inertial—observers). But as we stressed in chapter 3, the differences in both time and space measurements depend *only* on your inability to agree with Gertrude about what events are simultaneous, and this disagreement will obtain for any relative motion you and Gertrude undergo, uniform or accelerated. Accelerated motion will therefore cause the same *sort* of difference in temporal and spatial measurements as in uniform motion. The *amounts* of the particular differences will vary for inertial and accelerated observers, and, in fact, predicting the exact magnitudes of these effects for accelerated observers must be one of the goals of the general theory of relativity.

At this point Einstein's paper begins the formal mathematical development of his new theory of relativity (and of gravity). In section 5.9 we will return to say something about his mathematical approach—so unlike the one used in the development of special relativity—but, for the present, we can leave Einstein's paper because we have now discussed everything that we need to present the primary conclusions of the general theory.

Because accelerations lead to relativity of spatial and temporal measurements, if Einstein's principle of equivalence is correct the presence of mass (because it produces gravitation) should also affect time and length measurements. This particular consequence of general relativity theory seems very strange, and it is important to explain in some detail just how we arrive at it.

5.5 RELATIVITY OF TIME MEASUREMENTS

We argued in section 3.7 that as Gertrude's clock moves past you with a uniform speed, you will determine that it ticks at a rate different than you would measure were it at rest with respect to you. It is therefore not surprising that the same effect should be seen when the clock is moved past you with changing (accelerating) speed. But according to the principle of equivalence, the effects produced by gravitation must be the same as those produced by an acceleration, and so the mere presence of matter near Gertrude's clock will have the same effect as accelerating it; that is, the clock should slow down in the presence of matter even if the clock does not move past you at all.

This suggests an immediate experimental test of Einstein's principle of equivalence: hold a clock at rest near a large mass (for example, the earth); as gravity acts on the clock it should slow down. The stronger the gravity, the more slowly the clock should tick. The effect is very small, and so it had not been noticed prior to Einstein's prediction; but after the publication of the general theory of relativity attempts were made to verify the predicted "distortion" of time measurements. Since about 1960 instrumentation and experimental techniques have become sufficiently refined so that this "gravitational time dilation" effect can be measured in any well-equipped physics laboratory.

For example, clocks at the National Bureau of Standards in Boulder, Colorado, are about a mile farther from the center of the earth than clocks at the United States Naval Observatory in Washington, D.C. This difference in altitude means that the Boulder clocks experience weaker gravity than do the clocks in Washington, D.C., and so one might expect them to run slightly faster than the Washington clocks. The Boulder clocks are indeed *observed* to gain about fifteen billionths of a second each day relative to the Washington clocks, a result in quantitative agreement with Einstein's general relativity theory. But one need not even appeal to such differences in altitude as that between Boulder and Washington, D.C. R. V. Pound and G. A. Rebka in 1960 successfully measured the gravitational influence on time measurements using atomic clocks separated by a height of only 74 feet in a building on the campus of Harvard University.[18]

5.6 RELATIVITY OF SPACE MEASUREMENTS

Time is not the only physical measurement altered by an acceleration. According to special relativity, measurements of length (that is, measurements of space) are also affected by the uniform motion of objects. We would expect the same sort of effect in dealing with accelerated motions, and if the principle of equivalence holds, the presence of matter should also alter measurements of space. This effect should take place even if the object we are measuring does not move at all but is held at rest near a mass.

What exactly do we mean by the "relativity of space measurements?" Relativity of time measurements may be imagined easily since time is measured with clocks and we can observe changes in the rate at which various clocks tick. In a similar way, one can regard lengths as measured by applying various rules of geometry to a portion of space. For example, in

[18] The technical report of this experiment is given in R. V. Pound and G. A. Rebka, Jr., "Apparent Weight of Photons," *Physical Review Letters* 4 (1960): 337–341.

surveying a piece of land one usually applies the rules of plane geometry to measure the area. So a relativistic alteration in length measurements can be interpreted as a *change in the rules of geometry* applicable in a certain region of space. General relativity predicts that the rules of geometry will change in a region of space when matter is brought nearby.

We should first of all recognize that when the term ''geometry'' is used, one thinks of the subject studied in school, ''Euclidean'' geometry (based on Euclid's text written in the third century B.C.). That sort of geometry is characterized by a certain set of rules for solving problems—rules embodied in the axioms, postulates, theorems, and lemmas that constitute the basis of a grade-school geometry course. (For example, there is the familiar rule of Euclidean geometry that the area of a triangle is one half the length of the base times the height, or that the sum of the angles in any triangle is exactly 180 degrees.) If we think of space as characterized by such geometric rules in the same way that we think of time as characterized by clock readings, then according to Einstein's theory the presence of matter in space will alter the rules of geometry and so ''distort'' all of our measurements of space.[19] A triangle in a region of space near matter need not have 180 degrees as the sum of its angles, and the area bounded by the triangle need not be equal to half the product of its base and its height.

Let us illustrate these ideas with another example, and this time we will be introducing a mathematical term that is basic to general relativity theory. In geometry, the shortest distance between two points is a line called a ''geodesic.'' According to our schoolbook geometry, the shortest distance between two points is always a straight line, and so we would say that the geodesic in this system of geometry is always a straight line. But as we have already mentioned, the geometry we learned in school is not the only sort possible. In fact, it is not even the sort of geometry that best describes how things are on the surface of the earth.

How can a geodesic—the shortest distance between two points—be anything but a straight line? The fact that we ask such a question reveals how thoroughly we have been led to believe in the validity of the one sort of

[19] Relativistic descriptions of space and time have often been characterized as ''distortions.'' The word assumes one ''right'' set of geometrical rules (Euclid's, for example), or one standard clock for time measurements, or one normative structure to the cosmos from which the relativistic effects depart. But relativity theory asserts that no such rules, clock, or structure exist. Einstein's theories actually only ''distort'' from past preconceptions and idealizations that are less verifiably accurate than descriptions obtained by relativity itself. The words ''distort'' and ''distortion'' are thus used sparingly in this book and always in quotation marks to suggest their anachronism.

geometry taught in school. It seems so "obvious" that Euclidean geometry describes the geometric situation in our physical surroundings; yet this description is only approximate. Beginning in the late eighteenth century, mathematicians studied as scholarly exercises systems of geometry in which geodesics are not straight lines. These geometric systems are usually termed "non-Euclidean geometries," although they are sometimes also called "warped" or "curved" spaces. Certain non-Euclidean geometries were found to be completely consistent from a logical point of view, and so mathematically they appeared to be just as valid as Euclidean geometry. Einstein, through his work on relativity, ultimately showed that such non-Euclidean geometric systems also have bearing on models of the physical world.

We now want to explain what is meant by a non-Euclidean geometry. In particular it will be useful to ask just how we can visualize such geometric systems. But there are two sorts of problems facing anyone attempting to understand the properties of non-Euclidean geometries. First, we all are thoroughly convinced in school that there is just one valid sort of geometry: Euclidean geometry. Euclidean rules are so much a part of our thinking that it is difficult to imagine any other system. Second, it is extremely difficult to provide an accurate illustration of "curved" or non-Euclidean space since we ourselves are embedded in the very three-dimensional space whose "distortion" from the Euclidean norm we are trying to illustrate. This situation is analogous to that faced by philosophers, artists, and others who wish to discuss the nature of human existence. To study the human condition it is often desirable to "get outside" it to an extent, to step back somehow from the thing we are trying to study, to become an observer as well as a participant.

To suggest the character of non-Euclidean systems of geometry, it is customary in nontechnical treatments of general relativity to imagine ourselves removed from our surrounding three-dimensional space of length, width, and depth. To do this, we will create a fictional representation of the physical world.[20] Imagine that the world is not three-dimensional but two-dimensional, contained on a surface, characterized only by length and width

---

[20] It is worth mentioning that the fiction about to be developed has a long history in mathematical pedagogy as well as in science fiction. A classic account, written in the nineteenth century by a schoolteacher and intended for general readers, continues to delight people today: E. A. Abbott, *Flatland: A Romance of Many Dimensions* (New York: Dover Publications, Inc., 1952). For a more recent and challenging treatment, see Arthur C. Clarke, "The Wall of Darkness," in *The Nine Billion Names of God* (New York: Signet Books, New American Library, Inc., 1974), p. 94.

as is the "world" portrayed on a movie screen. You can think of yourself as a completely squashed, two-dimensional creature crawling in the surface world, which has no physical depth. It helps to imagine that you have been squashed to an infinitely thin piece of matter, rather like a little piece of tissue paper. You (the infinitely thin piece of paper) are free to move about your surface world, but you must remain *in* (actually part of) the surface.

Because we are really three-dimensional beings, we are able to step outside such a "surface world" in order to observe its properties. However, it is essential to remember that our fictional situation must be dealt with consistently if the analogy is to work. We repeat: you are an infinitely thin creature; there is only front and back and right and left in your world. There is no such thing as up and down.

First, consider the situation most familiar to all of us: a two-dimensional space that is flat, meaning a space with no "warp" or "curvature" (figure 5.5). This corresponds to the Euclidean geometry of our schoolbooks. A two-dimensional example of such a space is a plane. Figure 5.5 shows a finite section of what could actually be a flat surface extending infinitely in all directions, and within which you as a two-dimensional creature are constrained to live. Since your infinitely thin body must conform to the surface that is your space, viewed from our three-dimensional perspective you are not only infinitely thin, but you are also completely flat—part of the plane itself, in fact.

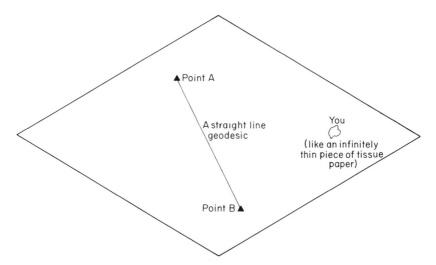

FIGURE 5.5

To see the nature of geodesics in this space, we have selected two points within it, A and B, and we have drawn a line between them following the shortest possible path. In the case of Euclidean geometry (or "zero-curvature" or "flat" geometry), such as one studies in school, the shortest distance between two points—a geodesic—is a straight line as shown. This example is not at all difficult to imagine because it is exactly the situation everyone deals with in grade-school geometry lessons. But to proceed we must tax the imagination a bit.

Figure 5.6 illustrates a two-dimensional "curved" space consisting of a "saddle-shaped" surface. Our presentation of this surface is complicated by the fact that this book is capable of rendering illustrations only on the surface of a page. You must imagine the saddle surface as it exists in three dimensions. Again, bear in mind that you are an infinitely thin creature (like a piece of tissue paper), but you are no longer flat. Your world—the surface—is no longer a plane. It is a curved surface world. And since you are constrained to exist within the curved surface, you must conform to the curve of your world, to your geometric fate. As a result you are somewhat "cupped." Figure 5.6 is again an illustration of a finite section of an infinitely large surface extending in all directions and over which you are free to move.

We have labeled two points in this world, A and B, and have drawn a line following the shortest possible path from one point to the other. This path represents a geodesic in this non-Euclidean space, and it is possible to

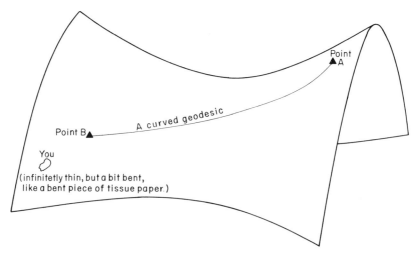

FIGURE 5.6

see that the geodesic is curved, corresponding to the fact that the space itself is curved.

As a final example of a curved two-dimensional space, consider the surface of a sphere (figure 5.7). Once again our illustration is confined to the
page of the book and so you must think of the surface as that of a three-
dimensional globe. This job of imagining is not nearly so difficult as with
the saddle-shaped surface. Because the globe of our earth is nearly spherical, we have been trained since childhood to think about spherical surfaces.
Indeed, we are rather sophisticated in our ability to conceptualize this example of a non-Euclidean geometry, although few school children are told
that that is what they are learning.

We have marked two arbitrary points, A and B, on this surface (for example, these points could represent two cities on the earth's nearly spherical surface), and we have drawn the shortest possible path between these
points. This line represents another curved geodesic. In the case of a spherical surface, the geodesic is given a special name: a ''great circle.'' Travelers occasionally hear this term in connection with flights between widely
separated cities on the curved surface of the earth. Because a great circle is
a geodesic, the great circle route represents the most efficient path to follow
on a trip over the surface of our approximately spherical earth.

While it is useful to imagine the world globe here, it is essential to bear

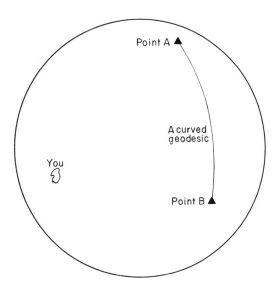

FIGURE 5.7a.

in mind that the two-dimensional *surface* of the globe is the world that you inhabit. In three dimensions, the surface route between the two points in figure 5.7a is not the shortest possible route between them. The shortest route for a three-dimensional being living on a three-dimensional sphere would be to burrow down through the earth from one point to the other, as shown in figure 5.7b. But you are not permitted to do this in your two-dimensional world because there is no "down into the earth" for you. You must confine all of your activities to the surface.

To this point our discussion has been about imaginary two-dimensional beings on imaginary two-dimensional surfaces. To understand the content of general relativity theory, one must apply the ideas illustrated in the fiction to our three-dimensional world. This represents an extremely difficult conceptual jump. Nevertheless, that is what mathematicians and physicists try to do as they work with non-Euclidean geometries.

When the original work on non-Euclidean geometries was done, curved spaces were probably considered by physicists to be purely mathematical exercises (if physicists considered this field of mathematics much at all). One of the first mathematicians involved in these studies was Carl Friedrich Gauss (1777–1855), who speculated on the possibility that Euclidean ge-

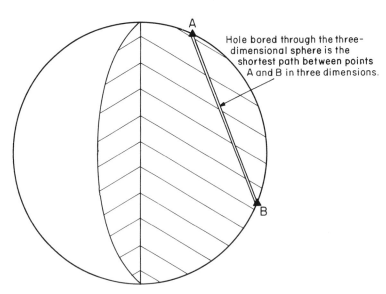

FIGURE 5.7b.

ometry is only an approximation to the geometric situation that actually describes the physical world; it is said, in what is probably an apocryphal story, that Gauss even attempted an experimental test of the rules of geometry. A large triangle on the earth's surface was carefully surveyed and was found to contain 180 degrees as required by Euclidean geometry.[21] A genuine and well-documented attempt at a similar sort of experiment, but on a much larger scale, was published in 1900 by the astronomer Karl Schwarzschild. Schwarzschild used observed star positions to establish a celestial triangle, and he then compared its properties to those of a Euclidean triangle. As far as he could tell the Euclidean description of space was valid.[22] But the motivation for such efforts in "experimental geometry" became of the utmost interest to physicists only after Einstein had created the general theory of relativity.

Einstein concluded that in fact non-Euclidean spaces can describe the geometric situation in nature. The mere presence of matter would affect the system of geometry applicable to a given region of space and so alter the shapes of geodesics there. For example, in one of these non-Euclidean spaces a triangle would not contain 180 degrees; the area included by the triangle's three sides would not be half the base times the height as it is in Euclidean geometry; the circumference of a circle would not be proportional to the radius of the circle, and so forth.

Einstein obtained equations that permitted him to predict the curvature of space, that is, the altered geodesics and rules of geometry that result from the influence of a given configuration of masses. In addition, he provided equations to predict how objects will move in such curved spaces (equations amounting to a "law of motion").[23] For many situations, Einstein's equations predicted essentially the same result obtained from Newton's models of motion and gravity. This corresponds to the fact that in most circumstances space is not appreciably "distorted" from the flat Euclidean geometry that we assume, and clocks are not usually dramatically

[21] This story is discussed in Arthur I. Miller, "The Myth of Gauss' Experiment on the Euclidean Nature of Physical Space," *Isis* 81 (1972): 345–348.

[22] Karl Schwarzschild, "Über das zulässige Krümmungsmass des Raumes," *Vierteljahrsschrift der astronomischen Gesellschaft* 35 (1900): 337–347. This paper was brought to our attention by H. P. Robertson, "Geometry as a Branch of Physics," in Schilpp, *Albert Einstein: Philosopher-Scientist*, I, 323.

[23] Although the equations may be easily applied *in principle* to solve for the curvature of space and the motions of objects, in practice the job is extremely difficult for all but the simplest situations. This is in marked contrast to the situation in Newtonian mechanics, where the equations are much more easily solved.

influenced by the sorts of masses we commonly encounter. For some situations, however, Einstein obtained answers that were quite different from Newton's, and when Einstein's predictions were compared with precise physical measurements, they turned out to be much more accurate than Newton's.

For example, in section 1.7 we discussed the limited ability of Newton's models of motion and gravity to describe accurately the orbital motion of the planet Mercury. In particular, remember that Mercury's orbit was found to rotate in space so that the point of closest approach between Mercury and the sun (the perihelion point) advanced each year by an amount that Newton's model could not predict. Mercury is the planet closest to the sun, and its proximity to that huge mass means that the relativistic effects on the nature of space will be greater than for any other planet. In his 1916 paper Einstein applied his newly formulated general relativity theory to calculate the motion of Mercury. To his great delight, he predicted the observed advance of Mercury's perihelion. "This discovery was, I believe, by far the strongest emotional experience in Einstein's scientific life, perhaps in all his life," asserts Abraham Pais, Einstein's colleague and recent scientific biographer:

> Nature had spoken to him. He had to be right. "For a few days, I was beside myself with joyous excitement." Later, he told Fokker that his discovery had given him palpitations of the heart. What he told de Haas is even more profoundly significant: when he saw that his calculations agreed with the unexplained astronomical observations, he had the feeling that something actually snapped in him.[24]

A second application of general relativity—and the one which, incidentally, brought Einstein his great popular fame—concerns the way that space influenced by the sun's mass alters observations of stars. Consider an observation made from the earth of a distant star (figure 5.8). We see the starlight approaching us from a certain direction in space. But when we view the same star some months later as shown in figure 5.9, so that the path of the star's light passes near the mass of the sun, we find, following Einstein's prediction, that the starlight comes to our telescope from a slightly different direction, demonstrating that the path followed by the light is "bent." As a result, the star's position appears to have shifted in the sky.

In this case, according to the general theory of relativity, the sun's mass

[24] Abraham Pais, *Subtle Is the Lord*, p. 253.

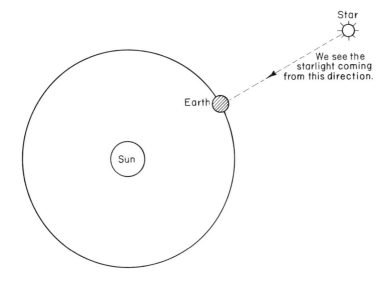

FIGURE 5.8

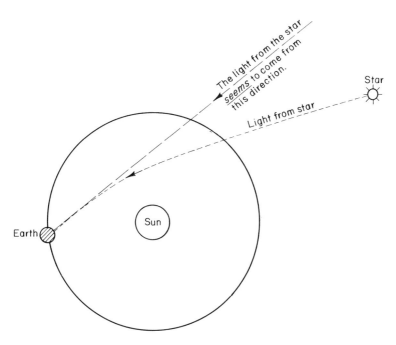

FIGURE 5.9

affects space, causing the light to follow a curved path, and the shift in the position of the star that astronomers observe is what Einstein predicted it should be. This shift is very small, so the effect was not noticed until Einstein's work prompted a careful search for it.[25] But its smallness is not the only problem with measuring this important prediction of Einstein's theory. The shift will be most pronounced (and hence most easily measured) when the starlight passes as near as possible to the sun as seen from earth. But observing stars that close to the sun is nearly impossible because the glare of the sun's light illuminates the earth's atmosphere to such an extent that the stars' much more feeble light is swamped.

There is a way around this difficulty, and it was recognized by Einstein himself in an early (1911) paper dealing with what would become his general relativity theory.

> As the fixed stars in the parts of the sky near the Sun are visible during total eclipses of the Sun, this consequence of the theory may be compared with experience . . . It would be a most desirable thing if astronomers would take up the question here raised.[26]

Astronomers did indeed "take up the question." An attempt to measure the effect in an eclipse visible from Brazil in 1912 failed because of poor weather (the bane of astronomy on earth); another eclipse was to be viewed from the Crimea in August of 1914 but failed because World War I broke out and caused the expedition to disband (it was, by the way, a German expedition en route in Russian territory when hostilities began); finally, British expeditions to Brazil and to Principe Island in May 1919 were able to confirm Einstein's prediction. Einstein wrote to his mother:

> Dear Mother, joyous news today. H. A. Lorentz telegraphed that the English expeditions have actually demonstrated the deflection of light from the sun.[27]

[25] The observed shift in the position of a star seen near the sun is about two seconds of arc. One second of arc is one-sixtieth of a minute of arc, which is, in turn, one-sixtieth of a degree of arc. One way to imagine such a small angular shift is this. Pablo stands at a railroad crossing holding up a quarter-dollar coin. When Gertrude views this coin with her unaided eye from a distance of *two miles*, the diameter of that coin will cover an angle of about 2 seconds of arc (if it could be seen at all by Gertrude).

[26] Einstein, *The Principle of Relativity*, p. 108.

[27] Pais, *Subtle Is the Lord*, p. 303. All of Pais's chapter 16 (from which this citation is taken) is nontechnical and covers in a most entertaining way the momentous events surrounding this major confirmation of Einstein's general relativity theory.

5.7 SPACE, TIME, AND SPACETIME

In our description of what general relativity predicts about the world, we have treated the relativity of space and of time measurements as separate effects; this is consistent with our discussion of special relativity theory in chapter 3. But in section 3.11 we pointed out that after Einstein's original work in 1905, Minkowski introduced the notion of a "four-dimensional space-time continuum," or, more simply, "spacetime." Physicists found special relativity much easier to conceptualize within the mathematical formalism of spacetime; physical situations in relativity could often be understood more easily by picturing them within the context of the spacetime diagram or Minkowski diagram (compare, for example, figures 3.28 and 3.29). The conceptual advantage of thinking in terms of spacetime is due to the fact that spacetime shows how the two entities, space and time, although distinct, are brought into a very close relation in special relativity.

The same sort of situation obtains in general relativity. One can separate the problem of effects due to the proximity of mass into temporal and spatial parts, but because these effects are so intimately related to one another, it is easier to think in terms of spacetime. In fact the general theory of relativity was developed entirely within this four-dimensional formalism (and that is how it is still presented in physics texts). The "warping" or "curvature" that we have been speaking about in terms of separate temporal and spatial effects is more simply and accurately viewed as the "warping" or "curvature" of a four-dimensional spacetime. Geodesics, then, are not the shortest possible distance between two points in space; geodesics refer to the *connections between events in spacetime*.

The idea of a geodesic as the shortest distance between two points in space is not difficult to understand, but now that we are speaking about geodesics in *spacetime* things are more remote from our ordinary experience and they are not as easily visualized. Also in chapter 3 we were dealing with Euclidean geometry and so all the rules of schoolbook geometry applied, even in the Minkowski diagrams representing four-dimensional spacetime. In particular, geodesics were straight lines. But in general relativity, the geometries are generally non-Euclidean, and so geodesics in our Minkowski diagrams will no longer be straight lines. Let's take a specific example of a geodesic in a Minkowski spacetime diagram and see how the four-dimensional picture compares with the more familiar picture of three-dimensional space and time as separate entities. We consider an object moving in space, first all alone, and then in proximity to another, larger mass like the sun.

Figure 5.10 shows the object moving in space from point A to point B when there is no other mass around. According to Newton's model of motion, the object will move in a straight line path (a geodesic in this Euclidean space) unless some force acts upon it. We can also represent this motion in a Minkowski diagram, as we did in chapter 3 (see figure 5.11).

As before, there is a space part of the motion (this is just the diagram shown in figure 5.10) and a time part of the motion (indicating the length of time required for the object to get from point A to point B in space). Both of these combine in the Minkowski diagram to produce the path in spacetime shown as the object moves from event A to event D. Notice, again, that the path in spacetime is a straight line, a geodesic in Euclidean space. So, another way of stating Newton's law of motion for this object is that the object moves along a geodesic in spacetime unless acted upon by a force.

Now let us change things by introducing the sun nearby our object. We know that masses near the sun can take up orbits about it, and let us suppose

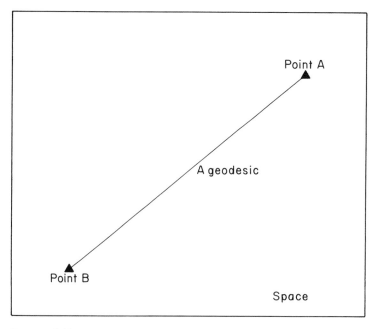

FIGURE 5.10

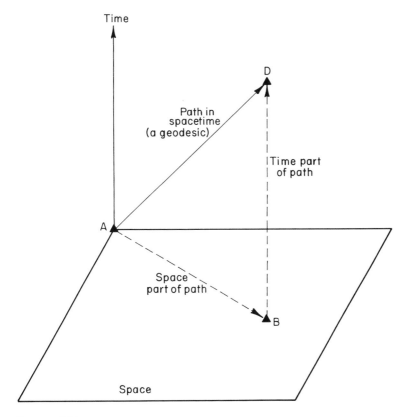

FIGURE 5.11

that things are appropriate for our small object to assume such an orbital motion about the sun; this situation is shown as a diagram of the space part of the motion in figure 5.12.

The object is no longer moving along a straight-line path. A Newtonian would interpret this fact in the following way: By Newton's first (or second) law, since the object does not move along a straight-line path, a force must be acting on it. As Newtonians, we further identify that force with gravity; we say that the gravitational pull of the sun on the object acts on the object continuously and deviates it continuously from a straight-line path (that is, from a geodesic in Euclidean space).

In contrast, here is what Einstein would say according to his general relativistic law of motion: "The object shown in figure 5.12 is moving along a geodesic in spacetime." Notice that this is just the same statement made

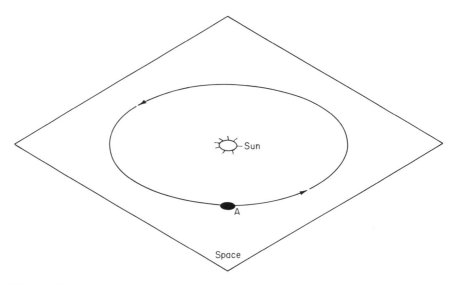

FIGURE 5.12

above by a Newtonian when the object was moving all alone in space (figure 5.10). Notice also that Einstein does not mention gravity at all in connection with figure 5.12. Relativity dispenses with the force that Newtonians call gravity. At first, Einstein's statement sounds absurd. Look at figure 5.12 again. The object starts out at position A in space, moves around the sun, and winds up back at position A again, and yet Einstein calls this "motion along a geodesic." If the motion is from position A back to position A, wouldn't the shortest possible path for the mass be just sitting at rest at position A? Why go all the way around the sun to follow a geodesic?

The answer is that we must deal not with the space part of the motion alone, but with the motion in spacetime. Einstein says that the mass follows a geodesic in spacetime, not just in space. Figure 5.13 is a spacetime diagram of the motion of our mass. With the full Minkowski diagram we can now see what is happening as the mass moves from point A in space, around the sun, and back to point A: as it moves in space the mass also advances in time. The path in spacetime is a helix, or "corkscrew." What before appeared to be a return to the same point A in space was actually a curved path from an event A to a very different event D in spacetime.

Remember Einstein's "law of motion." It asserts that the mass follows a geodesic in spacetime. Compare figure 5.11 with figure 5.13. When the

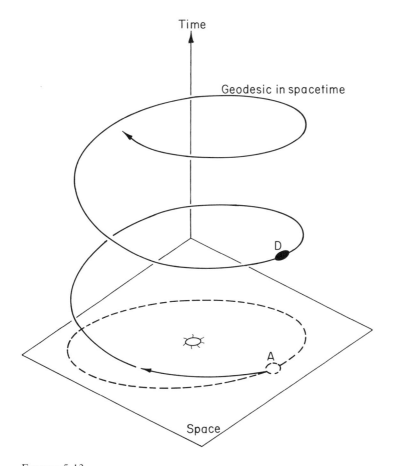

FIGURE 5.13

mass of the sun was absent, the geodesic of the object in spacetime was a straight line. With the mass of the sun present, the geodesic is now a curved line, corresponding to the fact that the presence of mass "warps," or "curves," spacetime. Thus we see that Newton, in dealing with the motion of things in space, and using time as a separate and unrelated entity, had to introduce a force to account for the behavior of one mass near another. Einstein uses the same statement (his law of motion) to describe the motion of both an object alone in space and an object near one or more other masses. This statement of Einstein's bears repeating: the mass moves along a geodesic. No gravitational force need be introduced to distinguish the two sit-

uations. The force of gravity, in general relativity, becomes a fictional consequence of the geometric properties of spacetime and is not a separate
phenomenon of nature the way it is in Newton's model of the world.

As we warned in section 5.6, Einstein's model of gravity still fails to
"explain" the phenomenon of gravity in the sense desired by Newton. Just
as Newton's model was unable to "explain why" masses should attract one
another, Einstein is unable to "explain why" masses should affect spacetime. He can describe the effect, but he provides no reason for its depend-
ence upon matter.[28] Einstein had hoped to elaborate his theory of gravity so
that all of the known forces of nature could be interpreted in an analogous
manner, that is, as geometric consequences of the behavior of spacetime.
He continued to search for such a "unified field theory" until his death in
1955.

5.8 The Shift of Star Positions and the Principle of Equivalence

In section 5.6 (see figures 5.8 and 5.9) we pointed out that because mass
influences spacetime, starlight will no longer follow straight-line paths near
the mass of the sun (the geodesics of light rays are no longer straight lines).
But according to the principle of equivalence, the bent path of a light ray
caused by a nearby mass should be equivalent to an effect produced by acceleration. It is helpful in clarifying both the ideas of spatial "distortion"
and the equivalence principle discussed in the preceding sections to see just
how this argument goes.

Consider the following (often-cited) thought-experiment carried out in
the rocket-powered laboratory we discussed earlier in connection with figures 5.1–5.3. Suppose a flashlight is mounted on one wall of the laboratory
(figure 5.14). When the rocket engine is turned off and the laboratory is not
accelerating, the beam of light moves straight across the laboratory and
strikes the opposite wall. We could fill the laboratory with smoke or dust in
order to render the beam visible as it crosses the width of the laboratory.
Now suppose that the rocket engine fires and the laboratory begins to accelerate (figure 5.15). A ray of light is shown leaving the lamp and heading
for the opposite wall. But as the ray moves across the laboratory, the speed

---

[28] The nature of matter (consisting of particles) in spacetime remained a puzzle whose solution Einstein hoped would provide insight into the meaning of certain results of the quantum theory of matter—in particular, the problems posed by the probabilistic descriptions of
phenomena that are inherent in quantum theory and which deeply disturbed Einstein.

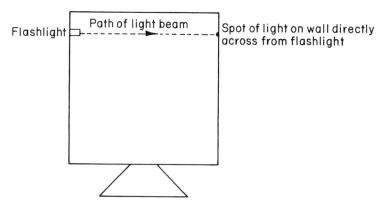

FIGURE 5.14

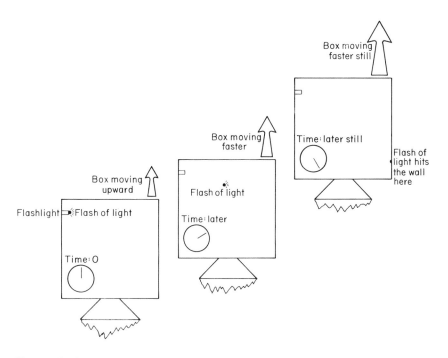

FIGURE 5.15

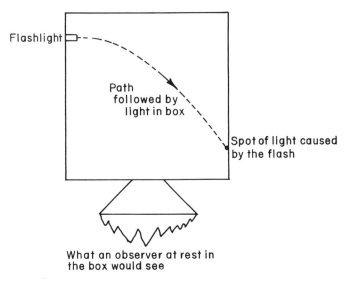

FIGURE 5.16

of the opposite wall continues to increase upward. The wall moves ahead of the light beam and so the beam strikes the wall, not at the spot directly opposite the lamp, but at a point some distance below that spot.

In fact, we find that an observer at rest in the accelerating laboratory would observe the ray following a curved path such as that shown in figure 5.16. This apparent bending of the light path is just the effect that Einstein predicted when the light ray moves near a mass such as that of the sun. That is, we would observe a path for the light like that in figure 5.16 if the laboratory were not accelerating at all but instead were at rest above a planet. As required by the principle of equivalence, the gravitational effect is the same as that produced by an acceleration.

5.9 THE QUANTITATIVE FORMULATION OF GENERAL RELATIVITY

Having presented the chief ideas underlying the general theory of relativity, we want finally to indicate the way in which the theory was formulated by Einstein. Armed with the principle of equivalence to take care of forces experienced when a frame accelerates, Einstein could be confident in stating his principle of relativity in mathematical terms:

> *The general laws of nature are to be expressed by equations which hold good for all systems of co-ordinates, that is, are co-variant.*[29]

In other words, it must be possible to express all laws of nature in such a

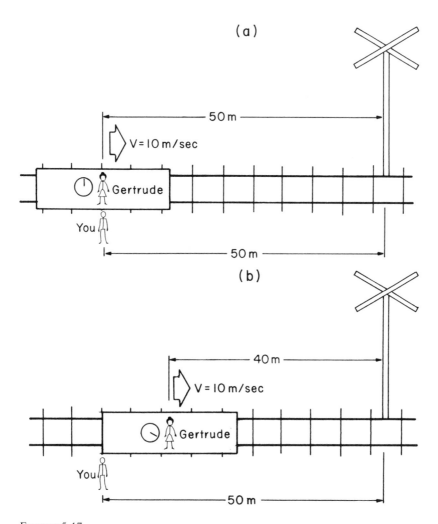

FIGURE 5.17

---

[29] Einstein, *The Principle of Relativity*, p. 117.

way that the equations representing those laws look the same in all frames of reference, accelerated or unaccelerated. The term ''covariant'' refers to this reference-frame independence of the laws.

By stating the laws of nature in this ''situation-independent'' form, it should be possible to see how different observers will disagree on measurements of quantities. The laws of physics will be the same for all frames of reference, but the numbers these laws yield will differ for observers in different frames of reference. The language here can get confusing; let's look at a specific example that we have discussed before.

Return to the railroad track. You are sitting on the embankment, and Gertrude, as usual, is riding her car along the track at a uniform speed of 10 meters per second with respect to you. Suppose Gertrude has a clock to mark the progress of her motion (figure 5.17). Also suppose that there is a railroad-crossing sign fixed to the embankment. You are able to describe the location of that sign in figure 5.17a by making a statement something like this: ''The sign is 50 meters from me.'' What will Gertrude say from her car? From her point of view, you and the sign are moving toward her at 10 meters per second. She therefore says something like this: ''At the instant shown in figure 5.17a (time zero on Gertrude's clock), the sign is 50 meters away from me and it is approaching me at 10 meters per second.''

Let's look at the situation after one second has elapsed on Gertrude's clock (figure 5.17b). From your vantage point the sign has not moved and so you would repeat the statement you made above: ''The sign is 50 meters from me.'' But Gertrude must now change her statement: ''At the instant shown in figure 5.17b (time reading of one second on Gertrude's clock), the sign is 40 meters from me, and still approaching at 10 meters per second.''

One could go on like this, drawing pictures and writing statements about the position of the sign at each instant of time, but there is an easier way of communicating the situations of both you and Gertrude. We can write two generalized statements, one for you and one for Gertrude, that will hold at all times.

For you:
distance of sign from me = 50 meters,
for Gertrude:
distance of sign from me = 50 meters minus my speed
multiplied by the time
elapsed on my
(Gertrude's) clock.

We can rewrite these two statements more briefly and abstractly like this:

For you:

$$D_{YOU} = 50 \text{ METERS},$$

for Gertrude:

$$D_{GERTRUDE} = 50 \text{ METERS} - VT,$$

where the symbol v stands for Gertrude's speed and the symbol T represents the time reading on her clock. The difference between these abstract statements and the pictorial or the verbal ones is basically that of brevity. We have replaced English words with other symbols, but the meaning is the same.

Now we have written two different statements about the distance of the sign for the two of you. At least the statements appear different at first glance, but notice that the symbols

$$D = 50 \text{ METERS}$$

of the two abstract statements are identical for the two of you. Taking advantage of this fact, we can arrange things so that your statement and Gertrude's are identical. We will have to be a little bit more abstract in our statement, but the result will be that the "law of motion of the sign" is the same for both your frames of reference; that is, we will have achieved a covariant statement of this "law." Here is the new statement:

> Distance of sign = 50 meters minus speed of sign with respect to observer multiplied by the elapsed time on the observer's clock,

or, even more briefly (and abstractly),

$$D = 50 - VT,$$

where D is the distance of the sign from the observer (that is, D can stand for either D_{YOU} or $D_{GERTRUDE}$),
  v is the speed of that observer with respect to the sign, and
  T is the time elapsed on the observer's clock.

Now, if we substitute your speed with respect to the sign in this statement of the "law" (that is, we let v = 0 in the equation), we find

$$D_{YOU} = 50 - 0 \times T$$

or

$$D_{YOU} = 50,$$

just as before. Similarly, for Gertrude we know that $v = 10$, and so

$$D_{GERTRUDE} = 50 - 10 \times T,$$

and when one second has elapsed on Gertrude's clock we have $T = 1$, and so we find

$$D_{GERTRUDE} = 50 - 10 \times 1 = 50 - 10 = 40$$

as before.

In other words, we have cast a situation that appears different for two observers into an abstract expression (an algebraic expression) that is identical for both observers. Each can use the same "law of motion" by substituting the appropriate (and different) value for a quantity v (the observer's speed relative to the sign) in the equation. Our law $D = 50 - vT$ may be termed "covariant" with respect to you and Gertrude.

The compactness of the algebraic form of the law is self-evident. What is not so evident is the fact that once framed in the language of algebra, the law may be altered by all of the rules of algebra and combined with other algebraic statements of physical and mathematical laws to create new laws. And if certain rules are followed, these new laws will also be covariant with respect to you and to Gertrude—that is, they will look exactly the same for the two of you. To compare how you and Gertrude will measure things, all we need to do is substitute the appropriate values for some of the symbols in the algebraically stated laws, just as we did for the law giving the distance from the sign.

Einstein's strategy in extending his special theory of relativity to a general theory was to find some mathematical device that would permit him to write the laws of physics in a covariant form, a form that would be identical for *any observer in any state of motion relative to another observer.*

Fortunately, mathematicians had created a device that was ready and waiting for Einstein when he needed it. In fact, Einstein was unaware of the necessary mathematics until 1912, when he appealed to his friend, the mathematician Marcel Grossmann, for help. It is said that Einstein pleaded, "Grossmann, you must help me or else I'll go crazy!" Grossmann worked with Einstein, tutoring him in the necessary mathematics and aiding him in his first steps at applying them to his developing general theory. The difficulty of the job is described by Einstein himself in a letter to a colleague:

At present [October 1912] I occupy myself exclusively with the problem of gravitation and now believe that I shall master all difficulties with the help of a friendly mathematician here. But one thing is certain, in all my life I have labored not nearly as hard, and I have become imbued with great respect for mathematics, the subtler part of which I had in my simple-mindedness regarded as pure luxury until now. Compared with this problem, the original relativity is child's play.[30]

In 1913 Grossmann and Einstein published a paper in which they began the development of the laws of physics in a language that was independent of a reference frame. These (covariant) statements of physical law are made in the language of what is called tensor calculus.

A tensor is a collection of numbers (or a mathematical formula standing for numbers) that represents some property (for example the mass or energy or the strength of electric and magnetic forces on an electrical charge) at each point in space. The space may have any number of dimensions (Einstein needed only the four dimensions of spacetime), and because the tensor represents some property of each point in space, a property *intrinsic* to that point in space, it is independent of the reference frame of any observer. By applying the appropriate rules of tensor calculus one can calculate from the given tensor the values of the corresponding property of space measured by different observers (such as you, Gertrude, Pablo, or Alice).[31] The trick is to formulate the laws of physics as tensor equations, because once a law is stated in the language of tensors, it will hold in any and all reference frames; results of specific measurements made by observers in any defined reference frame can then be extracted from the tensor equations as needed.

The general tensor description of non-Euclidean geometries had already

---

[30] Pais, *Subtle Is the Lord*, pp. 212, 216.

[31] This characterization of tensors is very rough. There are actually different types of tensors, each defined properly by rather difficult mathematical statements. We know of no really good discussion of tensors for general readers, and an attempt to provide one here would require an inordinate amount of text with a rather doubtful benefit; easily as much effort would be involved in explaining the mathematics as in explaining the underlying physical ideas that the mathematics serves to quantify. We can recommend one book that discusses general relativity for nonscientists in a quantitative way but without the use of tensors: Robert Geroch, *General Relativity from A to B* (Chicago: The University of Chicago Press, 1978). The approach in Geroch's book is graphical (using Minkowski diagrams), but many readers will find the book more difficult than our graphical discussion in appendixes A through C. At a more elementary level (but still using Minkowski diagrams) we can recommend W. J. Kaufmann III, *The Cosmic Frontiers of General Relativity* (Boston: Little, Brown and Company, 1977).

been worked out by mathematicians. The problem for Einstein was to cast a model of gravity in terms of tensors and in particular to describe in tensor equations the way in which matter affects the geometric properties of spacetime. How does one go about solving such a problem? The answer is that you obtain the tensor equations any way you can. Einstein manipulated existing tensor laws in combination with some ''inspired guessing.'' But the process was not in the least arbitrary, because he had definite criteria for judging when he had arrived at the correct formulation. The equations had to meet these tests: they had to predict what Newton's theory predicted for situations in which speeds are small compared to that of light and when masses are small (small departures of spacetime from the Euclidean case)—the so-called ''nonrelativistic limit''; when accelerations and masses were absent from the situation, the equations had to predict correctly the results of special relativity; and to these physical criteria Einstein added an aesthetic one: he sought equations ''which are *in their general covariant* formulation the *simplest ones possible.*''[32]

The actual process by which Einstein arrived at his final mathematical formulation of general relativity was lengthy. He finished in November 1915. The tensor equation describing the alteration of spacetime by matter which he then published is reproduced as (optional) box 5.2. Guided by his deep insight into nature, and relentlessly correcting his mistakes as he progressed, Einstein arrived at this brief but profound statement of the way in which material objects determine the structure of spacetime. Looking at this history, one gets the impression of success achieved by a stunning combination of dogged persistence, genius, and intellectual brute force.

ALL OF the predictions of general relativity that have been tested have been confirmed.[33] While Newton's theory, including his notion of a force of gravity acting between all pieces of matter in the universe, permitted him and subsequent Newtonian physicists to predict the way gravity influences the motions of objects, Einstein's theory of general relativity, with no *force* of gravity but with non-Euclidean spacetime, predicted these motions more accurately. In addition, Einstein's model comprehended new sorts of phenomena that Newton's model could not.

[32] Einstein, ''Autobiographical Notes,'' in Schilpp, *Albert Einstein: Philosopher-Scientist*, I, 69.

[33] A very recent discussion of the experimental tests of general relativity theory for nonscientists is given by Clifford M. Will, *Was Einstein Right?* (New York: Basic Books, Inc., 1986).

Box 5.2

Einstein's Tensor Field Equation of
General Relativity

These Greek letters above tensor symbols stand for the numbers 1, 2, 3, and 4, each representing one dimension of four-dimensional spacetime. They are used with the tensor symbols to indicate which dimension of spacetime the tensor is describing.

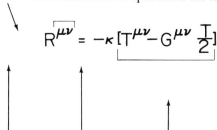

$$R^{\mu\nu} = -\kappa \left[T^{\mu\nu} - G^{\mu\nu} \frac{1}{2} \right]$$

This combination of tensor symbols describes the configuration of mass and energy that is at a given point in spacetime.

A constant of proportionality (see note 10 in section 5.3 for a discussion of this sort of number) allowing for the units of measurement chosen for quantities in the rest of the equation.

A tensor symbol describing the curvature of spacetime (the rules of geometry applicable at a given point of spacetime).

See section 5.9 for another discussion of tensors; appendix D.2 discusses the concept of a ''field.''

Special relativity in 1905 had created a highly counterintuitive vision of the nature of space and time; before that assault on ''common sense'' could be widely assimilated, general relativity continued the attack by supplanting Newtonian gravity, dethroning Euclidean geometry, and identifying additional bizarre properties of space and time. On the cosmic scale, too, Einstein's theory completed the job of displacing humanity from the center of the cosmos begun by Copernicus and Kepler. We are now prepared to complete the cosmological discussion begun in chapter 1.

6

COSMOLOGICAL CONSEQUENCES OF GENERAL RELATIVITY

> A universe so constituted would have, with respect to its gravitational field, no center.
>
> —A. Einstein

6.1 Introduction

We pointed out in chapter 1 that Aristotelian physics was used to support the geocentric model of the cosmos, a model locating human beings at or near the center of the universe. Later, Newtonian physics served to valorize the heliocentric model that displaced humans from their central location. But that heliocentric displacement was only partial: while the earth itself lost central status, the sun about which the earth orbits was thought to be at (or very near) the center of the astronomical universe. The great prestige of the heliocentric view persisted among most astronomers until the third decade of the twentieth century. Then a combination of telescopic discoveries and the application of Einstein's general theory of relativity to the structure of spacetime in the cosmos at large led to models requiring yet another centrifugal displacement of humans and their planet. Or perhaps it would be better to say that the new relativistic models made the ideas of centrality and displacement meaningless because the general relativistic universe is one without center.

In this chapter we discuss this postheliocentric vision of humanity's place in the cosmos. We begin by summarizing some key results of observational astronomy and the models that are currently used by astronomers to interpret these observations.[1]

---

[1] Other authors have handled the subject of relativistic cosmology (as well as other applications of general relativity including the notorious "black hole") with admirable clarity in works for nonscientists. We urge interested readers to pursue the consequences of general

6.2 OBSERVING THE UNIVERSE OF GALAXIES

Our sun and solar system exist within a vast stellar structure called the Milky Way Galaxy, represented schematically in figure 6.1. The sun is one of some 200 billion stars that are bound to this system by gravity. The detailed structure of the Milky Way Galaxy is quite complex, but for our purposes we can think of it as defined by a roughly spheroidal volume (called the "halo" by astronomers) some 600,000 trillion miles or more in diameter. The population of stars in this halo is generally quite sparse, so that if our galaxy were viewed with an optical telescope from some point far out in space the halo would not be at all the dominant feature. In our galaxy the stars are most densely crowded in a volume within the halo having the shape of a fat pancake with a large bulge in its center (see figure 6.1). It is this crowded volume (or "disk," to use the standard astronomical term) that is the most prominent visible feature of our stellar system. Seen face-on the disk would have a spiral or "pinwheel" appearance because the stars tend to concentrate along spiral streamers or "arms." Our sun is located in one of the spiral arms about two-thirds of the way from the center of the disk.

On a clear and moonless night, away from city lights (particularly in the summer), anyone can look up and view the disk of our galaxy as a band of light stretching across the sky. This band is called the Milky Way. Through a telescope the band of light is resolved into myriad stars comprising the disk of the galaxy. There are so many stars populating the volume of the disk that their light blends to the unaided eye and forms the Milky Way.[2] The sun in the disk of the galaxy moves once around the galactic center every 250 million years in a gravitationally bound orbit, just as the earth and other planets orbit the sun.

As a galaxy of stars the Milky Way is not at all unique. Observations made in all directions of the sky indicate that comparable galaxies of stars extend out billions of light-years into the depths of space. Some galaxies

---

relativity in some of these treatments, such as Timothy Ferris, *The Red Limit* (New York: Quill, 1983); Steven Weinberg, *The First Three Minutes* (New York: Basic Books, Inc., 1977); Kaufmann, *The Cosmic Frontiers of General Relativity*; and at a more advanced level—but still appropriate for the general reader—we highly recommend Gingerich, ed., *Cosmology Plus One*.

[2] If our view of the disk of our galaxy were completely unobstructed, we should see a very brilliant band of light stretching all around the sky. But our galaxy contains gas and "dust" particles as well as stars. These create a general "haze"; in addition, the gas and "dust" are sometimes collected into huge clouds between the stars. When viewed through these clouds and haze, the Milky Way takes on a rather blotchy appearance.

Face-on view of the Milky Way galaxy

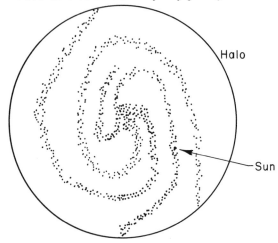

Side view of the Milky Way galaxy

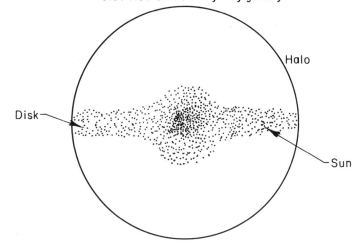

FIGURE 6.1

are much smaller than ours, some are much larger, some are spheroidal or elliptical in shape with no evident central disk or spiral structure, some are spiral pinwheels like our own Milky Way, and still others are of irregular shape. Figure 6.2 shows a collection of photographs of galaxies.[3] The first in this series shows a galaxy that is thought to be very much like our own seen "face on." A "side view" of our galaxy would be like that shown in the second photograph in the series.

In the last three paragraphs we have been mixing observations (for ex-

FIGURE 6.2a. The galaxy NGC 5457, thought to be similar to our own Milky Way galaxy seen face-on. The distance to this galaxy is 87 million trillion miles. The light that made this picture left the galaxy some ten million years ago.

[3] A more extensive set of photographs may be found in the book by Timothy Ferris, *Galaxies* (San Francisco: Sierra Club Books, 1980).

FIGURE 6.2b. The galaxy NGC 4594, thought to be similar to our own Milky Way galaxy seen nearly edge-on. The distance to this galaxy is 128 million trillion miles. The light that made this picture left the galaxy some fifteen million years ago.

ample, the observed objects called ''galaxies'') with the models used to interpret them (the galaxies are described as vast stellar systems comparable to the Milky Way). We must point out, however, that these models were not widely accepted until the 1920s. Although galaxies as patches of light in the sky were long known to exist, it was not known whether they were ''local'' objects actually part of the Milky Way or separate stellar systems (galaxies) comparable to the Milky Way. The detailed structure of the Milky Way and the sun's location within it were also subjects of lively debate. Most astronomers before about 1930 interpreted their observations by means of a model locating the sun near the center of the Milky Way star system, which, in turn, was thought to constitute the bulk of the stellar uni-

FIGURE 6.2c. A portion of the ''Coma cluster'' of galaxies. The distance to this cluster of galaxies is about 2,000 million trillion miles. The light that made this picture left the galaxies of this cluster some 220 million years ago.

verse; the objects now called galaxies were considered to be part of the Milky Way.

Just as Copernicus and Kepler displaced humanity from the center of the solar system, so astronomers in the twentieth century displaced the sun from the center of the universe and even from the center of our local galaxy of stars. Humans turning to astronomy for a vision of their cosmic situation then found themselves living on one of several planets circling an ordinary star, two-thirds of the way out from the center of an ordinary galaxy among a universe of billions of galaxies. Notice that each cosmological model supposing humans to be near the center of the universe eventually lost prestige. We will have more to say about this in section 6.6.

In 1929 the astronomer Edwin Hubble announced a discovery of fundamental importance to the science of cosmology. This discovery was not
about the location of the sun or Milky Way Galaxy among the other galaxies in the universe; rather, it concerned the motions of galaxies external
to our own. Hubble found that *distant galaxies all appear to be moving
away or "receding" from the Milky Way.* Furthermore, that outward motion of the galaxies has a pattern: *the more distant the galaxy, the faster it
appears to be moving away from us.* Speeds of recession up to 2,000 miles
per second were measured by Hubble and his associates. (Recent observations of very distant objects show some to be moving away at about 90 percent of the speed of light.) The observation that more distant galaxies move
away from us with greater speed is now known as "the Hubble relation."

The Hubble relation is often said to provide evidence that the universe is
"expanding." The word "expansion" is frequently misunderstood in this
cosmological context and its correct interpretation is a bit tricky as we will
explain in section 6.6. First, however, we turn to consider how Einstein's
general relativity theory comprehends the astronomical observations we
have been summarizing.

6.3 Cosmological Models from General Relativity

As we explained in chapter 5, one of the basic tenets of general relativity
is that matter determines the geometry of spacetime. This fact is also basic
to modern cosmological models that are founded on general relativity. This
point is so crucial to all that follows in this chapter that it bears repeating:
the matter in the universe determines the structure of spacetime. The spatial
component of spacetime defines the geometric rules that apply to surveying
various parts of the cosmos; the temporal component determines the history
of the universe. In particular, it is the matter in the universe that determines
the geometry of spacetime governing the large-scale motion of galaxies that
astronomers observe in the Hubble relation.

The usual cosmological analysis based on general relativity is carried out
in terms of a key number representing the average density of matter in the
universe.[4] Consider a cube one centimeter on each side; add up all of the
mass inside that cube—mass can be measured as the number of grams of

---

[4] One often sees the average density of matter in the universe symbolized by the Greek
letter *rho* with a zero subscript, ρ_0.

the matter present—and that number will be the average mass density (or just "the density") inside the cube. Density is defined to be the number of grams of matter (and all properties of the matter are irrelevant except for its mass) in a volume of space equivalent to the cube one centimeter on a side (that is, a cubic centimeter of volume). The measure of density is therefore the number of grams per cubic centimeter.[5]

The density of matter that we measure in space will obviously depend on the region of space in which we choose to locate the centimeter cube. A cubic centimeter of volume in the iron wall of the boiler on Gertrude's steam locomotive will contain a high mass, and so the mass density there will be correspondingly high. On the other hand, a cubic centimeter volume at some point in space midway between the moon and earth may contain only a few atoms of matter and will constitute an excellent vacuum by terrestrial laboratory standards; the density in this region of the universe will be very low.

But if we consider the density averaged over huge cubic volumes, millions of light-years on a side (one light-year is about six trillion miles), spread throughout the observable universe, then it is usually assumed (possibly incorrectly) that small-scale variations in density should be averaged out and the density of matter in one such cube should be very nearly that in any other such cube. The density values are still given in grams per cubic centimeter even though the volume over which the average is taken is much greater than one cubic centimeter. In cosmology, when reference is made to "the density of matter in the universe" one means the large-scale average mass density taken over cubic volumes millions of light-years in extent. As we point out in section 6.7, there are quite serious practical difficulties involved in actually determining a value for this average density.

We have said that general relativity specifies the geometry of spacetime for a given configuration of matter and that the relativity equations are solved for geometric properties in terms of the average mass density of the universe.[6] In the remainder of this section we will summarize some results

[5] Density describes how densely packed the atomic particles of matter are in a substance. Water in ordinary conditions has a density of about one gram per cubic centimeter; ice has a lower density (that is why ice floats in water). By contrast, the density of lead is about 11 grams per cubic centimeter and the density of air is .0013 grams per cubic centimeter.

[6] Our discussion here is a highly simplified and schematic sketch of the actual process. For one thing, cosmologists find it necessary to make certain simplifying assumptions in order to obtain practical solutions to the equations of general relativity. We have already mentioned one common assumption: that matter is uniformly distributed throughout space and has the same average properties wherever in space an observer is located. Such assumptions of ho-

of these calculations. Our discussion is simplified by the fact that the solutions that prove to be consistent with observed properties of the universe fall into one of three geometric categories; these categories are often termed "open," "Euclidean," and "closed" for reasons to be explained presently. The closed and open categories pertain to models of the cosmos in which, respectively, the average mass density is greater than or less than a certain value called "the critical mass density." The Euclidean category corresponds to models in which the actual mass density of the universe is exactly equal to the critical density.

According to cosmological models based on general relativity theory, if the average density of matter in the universe is just equal to the critical density,[7] then the predicted geometry of spacetime is Euclidean. Only for an average density equal to this critical value will the rules of high school geometry apply to the universe at large. It also turns out that such a universe will have an infinite amount of volume (like the infinite amount of room available to a two-dimensional creature exploring a Euclidean universe). The odds seem small that the universe would just happen to have an average mass density exactly equal to the critical value. It is far more likely that the actual mass density in the universe is greater or smaller than the critical value, in which case the geometry of spacetime falls into one of the other two possible categories.

If the average density is at all greater than the critical value, general relativity predicts that the rules of geometry in the cosmos at large will be non-Euclidean and the cosmological model is said to be "closed" because the amount of "room" available is finite, both in space and in time. On the other hand, if the average density is at all less than the critical value, the geometry is again non-Euclidean, but because in this case the amount of "room" in the universe turns out to be infinite, the cosmological model is

---

mogeneity may make the mathematical solutions simpler, but they are evidently not sustained by the observed properties of the astronomical universe. Observational evidence bearing on this question is difficult to obtain, and this point is under very active investigation now. Recent popular articles discussing some of this work are Dennis Overbye, "The Shadow Universe," *Discover* 6, no. 5 (May 1985): 12–25; still more recently and at a more advanced level, Jack O. Burns, "Very Large Structures in the Universe," *Scientific American* 255, no. 1 (July 1986): 38–47.

[7] The value of the critical density is not known with precision but is something like 0.0000000000000000000000000000001 (or in scientific notation, 10^{-30}) grams per cubic centimeter. Such a density is equivalent to having one atom of hydrogen (the least massive sort of atom known) in every two billion cubic centimeters of volume.

said to be ''open.''[8] The reason for the term ''critical density'' will now be clear; this single value for the average density of matter in the universe represents the dividing line between the two non-Euclidean categories of spacetime geometry: closed (finite) and open (infinite). We must remember that we are speaking here of the average properties of four-dimensional spacetime throughout the universe, neglecting local concentrations of mass (such as galaxies, planets, and people) that will cause local deviations of the geometric properties of spacetime from their cosmic average.

We must also be very careful in thinking about cosmologies in the ''closed'' category, those characterized by a finite amount of ''room.'' Just because the available room is finite does not mean that there is some cosmic ''border'' or ''edge'' marking a spatial boundary to the universe. Again we are dealing with a non-Euclidean concept that is difficult to imagine in three-dimensional space. A useful two-dimensional analogy is once again provided by the surface of the sphere. In such a two-dimensional cosmos a creature living *in* the surface has only a finite amount of territory available as she wanders about. But notice that although her world is finite (meaning a limited amount of room) it is also ''unbounded'' in the sense that as she travels she encounters no ''edge'' or ''wall'' marking the boundary of her cosmos. In three-dimensional space this ''finite but unbounded'' universe would correspond to a finite number of cubic miles available to travelers throughout the universe, but no boundary of any sort would ever be encountered marking the ''end'' of the available room. There would only be so much volume in existence and no more. In contrast, cosmologies with spacetime geometries belonging to the Euclidean or negative categories, there is an infinite amount of volume available in space (there is limitless ''room'').

Because the three categories of models refer to spacetime, there is also an inherent temporal component to each. Just as the spatial extent of a positive geometry is limited (a finite amount of volume available), so is its tem-

---

[8] The terms ''positive'' and ''negative'' are sometimes applied to closed and open cosmological models, respectively. This nomenclature results from the mathematical characterization of the geometric systems. One also frequently encounters a special notation for the average density of matter in the universe. Instead of using a value for the average density itself, some authors prefer to use the ratio of the actual average density to the critical density. This ratio is symbolized by the capital Greek letter omega, Ω. If Ω is one, the average density is exactly equal to the critical density, and the cosmological model is Euclidean. Similarly, if Ω is greater or less than one, the universe is non-Euclidean and closed or open, respectively. Recently, a theoretical Ω exactly equal to one has become fashionable.

poral extent (a finite amount of history exists). That is, a positive, four-dimensional spacetime had a definite beginning in time and it will have a
definite end in time. Before and after that beginning and end such a universe simply did not exist. Asking what happens "before" or "after" the
existence of this universe is meaningless. Space and time simply do not exist "before" or "after": there is no before or after. In contrast to positive
spacetime, cosmologies described by the negative and the Euclidean categories of geometry extend into the future indefinitely.

6.4 A NEWTONIAN VIEW OF RELATIVISTIC COSMOLOGICAL MODELS

The models of the cosmos we are discussing in this chapter are models
obtained from general relativity theory, but we can use our Newtonian
physical intuition to understand how the average mass density of the universe distinguishes the temporal and spatial characteristics of each category
of spacetime geometry. Consider the phenomenon of the Hubble expansion
from the standpoint of a Newtonian physicist. All of the galaxies are rushing away from one another. But it is known that the galaxies are massive
collections of stars and therefore they should exert mutual gravitational
forces on one another. The phenomenon of gravitation should act to retard
the separation. If there is only a small mass density in the universe, the
gravitational retardation will be slight and the galaxies will continue their
mutual recession forever. The volume of the universe will continue to grow
indefinitely and the cosmos will have no end temporally or spatially. These
are the properties discussed above in connection with the "open" category
of spacetime geometry.

On the other hand, if the average density of matter in the universe is very
great, the gravitational retardation acting on galaxies will be correspondingly strong and in time (the amount of time becoming shorter and shorter
as the mass density of the universe is greater—according to both the Newtonian and the general relativistic models) the gravitational pull of the matter of the universe will halt the increasing separation of the galaxies. Eventually the universe will stop "expanding." But gravity will continue to act
on the galaxies, tending to pull them back together. The galaxies then will
begin to move closer together again, and the universe will enter a "col-
lapsing" phase. But the "collapse" of the three-dimensional universe cannot go on forever. Eventually all of the matter in the cosmos will be compressed together, the collapse will stop, and at that time the present

universe will come to an end. The universe will therefore be of finite temporal duration. Such a universe will also have a finite extent in space (the growth of available room will be halted sooner or later by gravity, again in both the Newtonian and the general relativistic models). These are just the properties we have described for the closed category of spacetime.

One can imagine that the universe might "bounce back" from its collapsed state and begin a cosmic expansion all over again in a new birth event. One can even speculate that the universe repeats its cycle of expansion followed by contraction and birth over and over endlessly. Such a "cyclic universe" does have a certain aesthetic appeal because although each "life" of the universe is finite in duration, the process of cosmic birthing could recur indefinitely into the future.

We have used Newtonian ideas of gravitational force to describe how the average mass density of the universe can determine the geometric and temporal extent of the cosmos, and we have seen that small densities lead to a model in the "open" category, while large mass densities lead to models in the "closed" category. There will be some dividing line, according to this Newtonian view, between these two sorts of spacetime categories corresponding to a certain, "critical" value of average mass density.

Suppose we start with an open universe. The mass density is insufficient to bring the separation of the galaxies to a halt. Now imagine that somehow the mass density of this universe is increased. The gravitational pull of the matter in this universe on its galaxies will increase and the separating of the galaxies will be retarded, although the galaxies will still separate forever; we have only slowed and not halted the separation. If we continue to increase gradually the average mass density of the cosmos, the galaxies will still separate forever, but less and less rapidly. Eventually, if we increase the average mass density enough, we will reach a point at which the galaxies are just barely able to halt their own separation after a very long (essentially an infinite) time. The average mass density at which this occurs corresponds to what we have called "the critical density" and, as we have discussed previously, general relativity shows that the geometry of the universe then would be infinite and Euclidean.

THIS completes our summary of the sorts of cosmological models that are predicted using Einstein's theory of general relativity. We now return to the observed properties of the universe to see how these models comprehend them. First we will discuss Einstein's early attempt to apply general relativity to the observed astronomical "facts," and we will see how Einstein was led to make what he later called "the biggest blunder he ever made in his life."[9]

6.5 EINSTEIN'S BLUNDER: THE COSMOLOGICAL CONSTANT

Hubble published his discovery of the continuing separation of galaxies in 1929, and therefore it was unknown to Einstein when in 1917 he published the results of applying his new general theory of relativity to the structure of spacetime in the universe. Einstein's paper was called "Cosmological Considerations on the General Theory of Relativity."[10] The best astronomical evidence then available to Einstein suggested that on the average the matter of the universe was undergoing no systematic large-scale motion; accordingly, Einstein applied his general relativity equations to determine the structure of a universe he believed to be "at rest." Einstein recognized, of course, that on a small scale matter does undergo motion in the universe: planets move about the sun, and individual stars were known to move about in space. But his concern was not with the small-scale details of the universe. He was treating the average properties of huge samples of matter as we have in discussing the average density of matter in the universe. He neglected any local motion of small individual objects like stars or planets and assumed that the matter of the universe on average was at rest or, to use Einstein's term that recognizes that his assumption is only an approximation, "quasi-static." Curiously, his relativity equations predicted that the matter of the universe should not be standing still but instead undergoing a systematic motion. To force his theory into conformity with his understanding of the large-scale static condition of matter in the universe, Einstein introduced a new term into his general relativity equations, a term multiplied by something since called the "cosmological constant." As Einstein says in his paper, "That term is necessary only for the purpose of making possible a quasi-static distribution of matter, as required by the fact of the small velocities of the stars."[11]

This decision of Einstein's was unfortunate. For one thing, in 1922 the

[9] This is reported to have been Einstein's comment to George Gamow and may be found in Gamow's autobiography, *My World Line* (New York: The Viking Press, 1970), p. 44. The comment is also mentioned in Gamow's article "The Evolutionary Universe," in Gingerich, ed., *Cosmology Plus One*, p. 16.

[10] This paper appears in translation in *The Principle of Relativity*, pp. 177–188.

[11] Einstein, *The Principle of Relativity*, p. 188. Modern students of physics might put it another way: Einstein "fudged" his equations to make the answer come out the way he thought it should.

mathematician Alexander Friedman found that Einstein had done something all high school students of algebra are warned never to do. He had divided both sides of an equation by a term that could become zero. In fact, even with his ad hoc term including the cosmological constant, the equations predicted that there should be a systematic motion of matter in the universe; the matter could be flying apart in all directions, or the matter could be "falling" in on itself. Einstein's addition of the term to his relativity equations in an attempt to make the universe "stand still" had been in vain. Later, of course, another aspect to this "blunder" became evident with the recognition of the Hubble relation. In fact, the large-scale distribution of matter in the universe is undergoing a systematic motion, one of "expansion." The irony is that Einstein in this case did not listen to the message of his own equations. Had he trusted his general theory more, he might have predicted the expansion of the universe before it had been observed (as he so successfully predicted many relativity effects years before they were observed).

6.6 INTERPRETING THE HUBBLE RELATION AS AN "EXPANSION"

In section 6.2 we cautioned against giving a simple interpretation to the "expansion" of the universe suggested by the Hubble relation. Having explored some properties of the cosmological models predicted by general relativity, we must now explain the meaning of that caution. When we understand that the Hubble relation is an observed separation of galaxies in all directions of the sky, how do we imagine this phenomenon? Surely the most simple interpretation is that shown in figure 6.3. All of the galaxies in the universe are rushing away from our Milky Way galaxy, moving out into the preexisting void of intergalactic space. In other words, we are at the very center of a cosmic "explosion." Such a model is straightforward. The evidence provided by the Hubble relation seems compelling. But this model is also deeply disturbing to cosmologists who, throughout the history of astronomy, have been proven wrong whenever they assumed that humans are anywhere near the center of the astronomical cosmos. But how could one possibly interpret the Hubble relation and at the same time keep humanity away from "the center" of things? We *observe* the galaxies rushing away from us in every direction.

General relativity does provide a way of interpreting the Hubble relation that does not suppose humanity to be anywhere near the center of the cos-

The Milky Way galaxy

FIGURE 6.3

mos. In fact, nothing is near the center of the cosmos according to these models because there is no such thing as a center to the relativistic cosmos. Imagine once again that you are a two-dimensional creature living in the surface of a sphere. That is, your cosmos is the spherical surface. Locate yourself in a two-dimensional "milky way" galaxy of stars in that surface as sketched in figure 6.4a. But you must not picture this spherical surface as that of an ordinary globe. That would be decidedly prerelativistic. The spherical surface that we must use in relativity is able to stretch; you can think of the surface as made of rubber—in fact, you can think of it as the surface of a perfectly spherical balloon. Now let the spherical surface expand—start to blow up the spherical balloon, in other words. This is suggested in figures 6.4b and 6.4c. As the size of the surface grows (as the balloon blows up) the galaxies separate from one another. To an astronomer living on your galaxy all of the other galaxies would appear to be rushing away from yours; the expansion of the universe would appear to be centered on you. But now imagine that you are living on any other galaxy in the spherical surface. You would see the same sort of expansion away from you. In other words, in this two-dimensional universe, the separation of galaxies appears the same no matter where in the universe one is located.[12]

_____

[12] If this paragraph and the discussion to follow become difficult to visualize, it will be worthwhile for the reader to take a (reasonably round) balloon and partly inflate it. Put little pieces of masking tape on the balloon surface to represent galaxies (you may want to mark your home galaxy with a red dot to distinguish it from the others), and then slowly inflate the balloon. Study the apparent motions of the other galaxies from the imagined perspective of

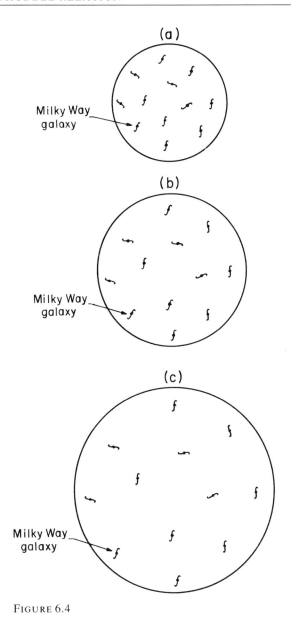

FIGURE 6.4

an astronomer who lives on your galaxy as the inflation proceeds. Then move your vantage point to another masking-tape galaxy and imagine how galaxy motions appear there. By the way, masking-tape galaxies are much better than drawing dots on the balloon surface to represent galaxies because drawn dots would expand along with the rubber sphere, implying that

There is no one "center" from which the separation takes place. Another way to put it sounds paradoxical but it is true: *every* point in the universe is a "center" of the expansion. That is, the expansion takes place from every point on the surface, uniformly, over the entire surface.

The analogy we have been using involves two-dimensional observers living in a surface universe. As in our discussion of contrasting geometries in section 5.6 we must now ask that you imagine the same sort of process in three dimensions. An analogy sometimes quoted to help with this three-dimensional visualization is provided by raisin bread rising in a hot oven. The raisins represent the galaxies and the dough represents the vacuum of space. As the dough "rises" (or more accurately "expands" into the surrounding space) the galaxies (the raisins) all move away from one another. We mention this analogy because it helps in making the transition from our image of expansion in two dimensions to the analogous process in three dimensions, but we hasten to caution that the "raisin bread" analogy is highly misleading in two respects. First, the raisins near the sides of the baking dish are privileged. Looking out in one direction into space (into the dough) observers on these raisin galaxies see other galaxies; looking out in another direction they see the side of the dish. This is not at all the situation we are here attempting to describe in the relativistic view of the universe. There is no "edge" or "side" to the relativistic cosmos any more than there is an edge or boundary in the balloon-surface world we discussed earlier.

The second difficulty with the "raisin bread" analogy is that the raisin bread grows in volume by expanding into a preexisting volume in the oven—that is, there is a void that the dough expands to fill. This is not the situation in the expanding universe of general relativity. Once again, return to the two-dimensional (balloon surface) analogy. The only "room" available is the surface of the balloon, and that surface area is growing as the balloon inflates and the galaxies move apart. As a three-dimensional being you see the surface expanding outward into a third geometric dimension, but you must be consistently two-dimensional in your perspective if the analogy is to work. From the perspective of a two-dimensional creature in the surface, the amount of room available in the universe continues to grow, and the growth takes place at all points in space. Furthermore the growth is *not* "into" any preexisting room. More room (or volume in the

---

the size of the galaxies changes as the balloon (the universe) expands. This is not correct. The sizes of the galaxies stay fixed.

three-dimensional case) comes into being at all points in space and the galaxies separate from one another as they "ride along" with the growing available room in the universe.

The Hubble relation also has profound consequences for our understanding of the temporal component of spacetime, the life of the universe. The galaxies are now observed to be rushing away from one another; this means that minutes or years or eons in the past they must have been closer together. In fact, if we assume that the increasing separation of the galaxies we observe now has always been a property of the universe (and this is consistent with general relativity theory), then at some time far enough in the past all of the matter in the universe must have been compressed together in an unimaginably dense state in some vanishingly small volume of space. That time would mark the "birth"[13] of the universe. In other words, the fact of the Hubble relation can lead us to the conclusion that the universe (as the science of cosmology defines it) had a beginning.[14]

6.7 WHAT TYPE OF UNIVERSE IS THIS?

In section 6.3 we cataloged the possible cosmological models provided by general relativity theory. The most interesting question is, Which of these possibilities corresponds to the universe in which we actually live? The theoretical models are distinguished by the value of the average density of matter in the universe. One straightforward way to determine which model most closely corresponds to our universe would be to measure the average density of matter. Unfortunately this approach does not work. In our discussion of the observed properties of the universe we focused our attention on galaxies of stars because it was through observations of these stellar systems that the Hubble relation and the "expansion" of the universe were first discovered. It was even thought at one time that galaxies and groupings of galaxies represent the primary "building blocks" of the

[13] Other terms commonly applied to this event are "origin," "creation event," and "big bang."

[14] But a beginning for the universe is not a necessary conclusion to be drawn from the Hubble relation. In fact, the so-called "steady state" model is consistent with both the observed Hubble relation and with general relativity theory, yet it describes a universe with no beginning or end in time. There is now, however, good observational evidence (independent of the Hubble relation) that the universe did have an origin in time of the type just described. Radio astronomers are able to observe electromagnetic radiation coming from all directions in space that has just the properties predicted for the vestigial effects of the cosmic "birth." For this reason (and for others as well) the steady-state universe is not favored by cosmologists today.

cosmos. But for some years evidence has been accumulating that points to some other component of mass—as yet invisible but detectable through its gravitational influence—as a significant, perhaps even the dominant, constituent of matter in the universe. Some research has suggested that the visible matter in galaxies may represent only about 10 percent or even less of the matter in the cosmos.[15] We emphasize that these new discoveries in no way alter either the general theory of relativity or our discussion of its cosmological ramifications; this newly discovered matter in the cosmos, if it does exist in the quantities suggested, only changes the dominant agent responsible for the geometry of spacetime. But to measure the average mass density in the universe by direct observation requires that we add in all of the matter in a given large volume of space, and therefore all of the matter must be detectable somehow and its quantity must be determinable from observations. Unfortunately our knowledge of the suggested "invisible matter" is as yet too poor to permit a good estimate of the average mass density (although some recent estimates would clearly put the average mass density in excess of the critical mass density, leading to the conclusion that the universe is closed). And so, while it is the average density of matter in the universe that determines the cosmological model, other observed quantities must be used to infer the average density indirectly, at least at this stage in the history of astronomical research.

We mentioned in section 5.6 that in 1900 the astronomer Karl Schwarzschild attempted to determine whether or not Euclidean geometry was valid over astronomical distances by interpreting observations of stars. Such a determination of geometric rules on an astronomical scale would suffice to estimate the average density of matter in the region of space surveyed. But the observational problems are formidable. Space turns out to be so nearly Euclidean that any departures in measured geometric properties are much too small to be detected with the observational techniques available to Schwarzschild or even to astronomers at present. Schwarzschild could only conclude from his data that any departure from Euclidean geometry was not detectable. This attempt to determine the geometry of space was years in advance of Einstein's theory. It was motivated by curiosity about the possible physical significance of the non-Euclidean geometries that had been invented by mathematicians. Since publication of Einstein's theory of relativity and the attendant cosmological models based on

---

[15] For a recent popular summary of this work, see Dennis Overbye, "The Shadow Universe."

non-Euclidean spacetime, cosmologists have made strenuous attempts to determine which model of the universe best fits astronomical observations.

A comprehensive review of the various techniques that have been proposed to determine the best cosmological model would require an inappropriately long discussion. To illustrate the sort of approach that can be taken we will here mention just one of these indirect approaches.[16]

We have already remarked that the greater the average mass density in the universe, the greater the gravitational force tending to pull the separating galaxies back together. It is therefore reasonable that there should be a close connection between the mass density and the rate at which the galaxies are slowing (or "decelerating") their recession. In principle, if one could measure the slowing of the galaxies, one could calculate the average mass density of the universe. This is a valid approach in principle. But how in practice does one measure the slowing of galaxy recession? Such an observation might be made if time travel were possible. One could measure the rate at which two galaxies are separating today and then go back in time to measure the rate at which the same two galaxies were separating long ago. The effects of the slowing down would become apparent if one went back something like a billion years into the past. But how does one travel back in time to carry out this observation?

Fortunately, time travel of a sort is common in observational astronomy because of the finite speed of light. When a photograph or any other sort of observation is made of a distant galaxy (like those shown in figure 6.2), the light that is collected from the galaxy to make the picture left its source long ago; the more distant the galaxy, the longer ago the light must have left it. The picture that we now receive of that galaxy is not of the galaxy today but of the galaxy as it was as long ago as it took the light to reach us.

The same light that we use to take a picture of a galaxy can also be used to measure the speed with which that galaxy is rushing away from us.[17] As-

---

[16] An excellent review of various techniques is given by J. Richard Gott III, James E. Gunn, David N. Schramm, and Beatrice M. Tinsley, "Will the Universe Expand Forever," in Gingerich, ed., *Cosmology Plus One*, pp. 82–93.

[17] One uses the Doppler effect. This is commonly encountered while listening to sound waves in certain situations. Imagine that you stand by a road and a truck passes you while sounding its horn. As the truck approaches you and then passes, the pitch of the horn varies: the pitch is higher when the truck is approaching, and then it suddenly falls as the truck passes and recedes down the road. The pitch of sound is determined by the number of sound waves that strike your eardrum each second. In the case of light, the "pitch" of the waves corresponds to color. The higher the pitch, the more blue the color. So if an observer and a source of light are approaching one another, the observer encounters more light waves each second

tronomers have also devised various techniques for estimating distances to galaxies; in fact, it is just this combination of speed and distance measurements that is used in determining the Hubble relation. But if more and more remote galaxies are observed in order to determine their distance and speed, the light received must have left the galaxies longer and longer ago, and so the Hubble relation we observe for very distant galaxies should differ from the one we obtain from those nearby. In particular, the more distant galaxies (seen in more remote times) should be separating faster than suggested by the Hubble relation for nearby galaxies, because longer ago the mutual gravitational attraction of the galaxies would not have had as much time to slow down the recession.

By comparing the Hubble relation for very distant galaxies with that for galaxies nearby, one should see the effects of gravitational retardation and be able to infer the value of the average density of matter in the universe. This is a legitimate observational protocol for determining the best cosmological model from general relativity. Unfortunately, there are two practical problems with its implementation. First, very distant galaxies are also very faint and therefore extremely difficult to observe. Second, the indicators used to determine distances to remote galaxies are not sufficiently reliable to obtain accurate distance values.

The result is that this technique has not yet succeeded in determining the density of matter in the universe. The same can be said for all of the other indirect techniques available for determining the average density. The observational values are just not accurate enough to discriminate between the three possible categories of cosmological models. New observational techniques are being developed and the advent of space astronomy (which removes telescopes from the highly absorbent blanket of the earth's atmosphere) gives cause for some optimism that the "cosmological problem" may well be solved before too long.

In the meantime, we are left to contemplate what it means to live in a finite but unbounded universe that one day will collapse back on itself, or

---

and the "pitch goes up," meaning that the light becomes more blue; if the source and the observer are separating or "receding" from one another, the observer encounters fewer waves each second, the pitch of the light waves falls, and the light looks more red. By breaking the light into its component colors—in a device called a spectrograph—the shift in color of the light from a galaxy can be measured and the galaxy's speed calculated from a simple formula. It is important to emphasize that it is the *pitch* (or, to use the physicists' term, the "frequency") of the waves that depends on how the source and the observer of the light are moving with respect to one another. The *speed* of the light waves is the same no matter how the source and the observer move with respect to one another.

an infinite universe that will go on forever, or perhaps a Euclidean universe precariously balanced between open and closed—possibilities that have all been made accessible to observational confirmation by general relativity. But in providing us these possible cosmological models, Einstein's theory has also deprived us of an important organizing principle. The Copernican Revolution displaced our planet from the center of the cosmos, and by the 1930s astronomers had displaced our sun from the center. General relativity did away with the very "fact" of a center to the cosmos. It is one sort of revolution that creates a centrifugal shift in human perspective, a shift that sets us adrift in space so that, as John Donne said, "The Sun is lost and th'earth and no man's wit / Can well direct him where to looke for it." It is quite another sort of revolution that renders obsolete *any* orientation with respect to a cosmic center. But this is a subject for another book.

WE CLOSE this chapter with a word of caution. Because the techniques for deciding the best cosmological model within the context of general relativity theory are perhaps at hand, and because the question of the nature of the universe is so exciting, work in this field is vigorous. We have already suggested that the answer may be near. But answers are suggested by research all the time. Sometimes these findings appear in articles for nonscientists. And sometimes the same answers are suggested by several different research groups working with different techniques. When different approaches point to the same conclusion (for example, the universe being open or closed), the temptation can be strong to turn in the direction of the herd. We noted earlier the outstanding article by Gott, Gunn, Schramm, and Tinsley, "Will The Universe Expand Forever?" in *Cosmology Plus One*. These authors point to evidence from many different fields, all suggesting an affirmative answer to the question raised in the title of their paper: they conclude that the universe is open and infinite. It is probably fair to say that this was the "best guess" up until a few years ago. Then a variety of arguments—theoretical and observational—seemed to turn opinion. By 1985 Euclidean or even closed models seemed to be popular again. The suggestion we extract from this is simple: readers of science articles should regard any "definitive answers" to the cosmological problem with healthy skepticism, even when evidence from various sources is consistent. That is one consequence of living in a vigorous and exciting time of scientific research, research initiated by Einstein's general theory of relativity.

CODA

After a little while I murmured to Picasso that I like his por-
trait of Gertrude Stein. Yes, he said, everybody says that she
does not look like it but that does not make any difference,
she will, he said.

—Gertrude Stein in
The Autobiography of Alice B. Toklas

WE epigraphed our book with Franz Kafka's parable of the leopards in the
temple because it suggests a fundamental process of science: bizarre phenomena are ultimately assimilated into the domain of validity of available
scientific models. At the same time, the parable suggests a cultural consequence of Einstein's work. In creating new models to comprehend certain
leopards in the temple of physical science, he tempted new ones to enter
our culture at large: a centerless universe that may be finite but unbounded;
time and distance scales that depend on the observer's state of motion;
masses that grow with relative speed. Now these, too, are becoming a part
of our ceremony.

Appendix A

The Lorentz Transformations and Minkowski Diagrams

A.1 Introduction

In section 3.11 we discussed the representation of events on a Minkowski diagram. We remarked that once an inertial observer has used her own measurements of distance and time to locate an event on such a diagram, measurements for the same event made by any other inertial observer may be found without using the Lorentz transformation equations. It is the purpose of appendixes A and B to show how this is accomplished. In appendix A we illustrate, without using algebra, the rules for solving the Lorentz transformation equations on a Minkowski diagram. We also provide justification for the conditions discussed in section 4.8 in which a temporal sequence of events may appear reversed to different observers. Appendix B uses algebra to justify the rules for solving the Lorentz transformations with Minkowski diagrams. Appendix C, which discusses particles that move faster than light, depends on appendix A but not on appendix B.

We begin with a review of Minkowski diagrams in somewhat more detailed and quantitative terms than in our original discussion.

A.2 A Review of Minkowski Diagrams: Still More Experiments on the Railway Track

We will again consider events taking place along the railroad track invoked so often in this book. Suppose that the track has markers fixed to the rails to indicate distance from some arbitrary benchmark on the track (figure A.1). These markers will permit us to measure the spatial location of events that happen on the track. (We are dealing with a one-dimensional spatial situation in order to simplify the discussion; an elaboration to two or three dimensions would be straightforward.)

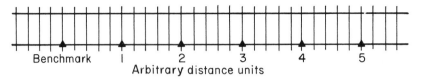

Figure A.2 shows you standing at the benchmark and Pablo at rest on the 6 distance-unit mark. You and Pablo have digital alarm watches synchronized to the common system of time; at exactly noon on the common system your watches beep and Pablo rolls a small hand car along the track toward you. At 4 seconds after noon on your watch, the hand car hits your foot. We therefore can discuss the three events shown in table A.1. These events are located in figure A.3, a diagram of the track showing only a side (ground-level) view of things. Event C falls right on top of event A, meaning that the two events happen at the same point on the track (the same point in space). But events A and C are not simultaneous to you, and the diagram is therefore somewhat misleading since it seems to suggest that the two events are coincident in every respect. Notice too that the very meaning of the word "event" involves a specification of "what," "when," and "where." We have differentiated "what" by giving the events different letter designations. Events A and C took place at the same point in space (the "where"). It remains to specify the difference in time of the events. To portray this difference (and, in general, to portray the time specification

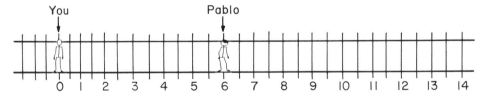

FIGURE A.2

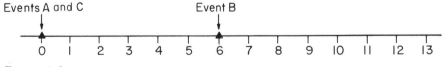

FIGURE A.3

TABLE A.1

Event	Common Time	Place	Description
A	12:00	0	Your watch beeps
B	12:00	6	Pablo rolls car toward you
C	12:04	0	Car hits your foot

of any event) we use a ''time line.'' One occasionally sees these in history books: various events are located along a line to indicate how they are related to one another in time. Figure A.4 gives an example. The same scheme can be used to represent our events A, B, and C (figure A.5). Just as we locate the place of occurrence of an event on the track by showing the number of distance units from the benchmark to the event on a horizontal line, so we locate when an event happens on a vertical line that passes through the benchmark in figure A.6. The numbers on this time line, or ''time axis'' as it is usually called, are time units on the common system. The three events specified in table A.1 are portrayed in figure A.7, where we have added some symbols to make the drawing clearer. ''T'' stands for the time reading on the common system and ''D'' stands for the distance that you measure from the benchmark. The distinction between events A and C is now clear, in contrast to figure A.3.

So far our discussion has been very close to that of section 3.11, albeit with more quantitative detail. Before we proceed to discuss the Lorentz

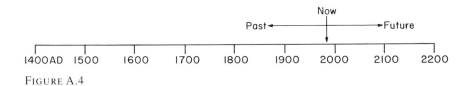

FIGURE A.4

FIGURE A.5

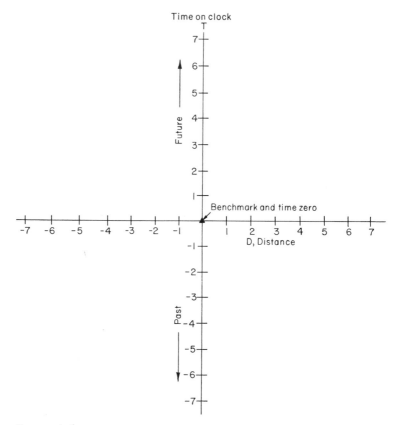

FIGURE A.6

transformations and some of their consequences as illustrated in the Minkowski diagram, we must first present examples of the way that several important types of phenomena appear in these diagrams. Use will be made of
these examples later in this appendix as well as in the following two.

First, let us return to the Minkowski diagram describing the events A, B,
and C that we have just discussed. We can be even more specific about the
circumstances surrounding these events, for we can use this sort of diagram
to represent the detailed motion of the hand car along the track. In figure
A.8 the broken line represents the location of the car (D) at each corresponding moment of time (T). As we remarked in section 3.11, points and
lines representing events in these diagrams are called ''world points'' and
''world lines.''

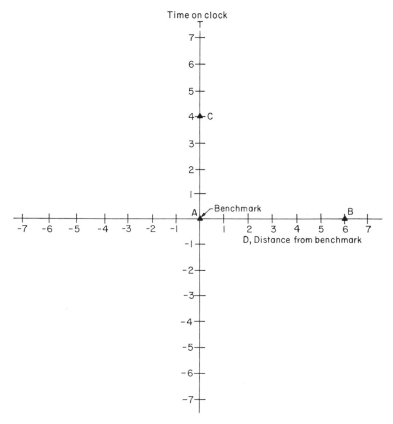

FIGURE A.7

You could roll a car to Pablo in the sequence of events listed in table A.2
and represented graphically in figure A.9. Or, you could roll the car away
from Pablo as shown in figure A.10.

Just as the world point of the car follows a path in a Minkowski diagram

TABLE A.2

Event	Common Time	Position	Description
A	0	0	You roll a car to Pablo
B	7	6	Car hits Pablo's foot

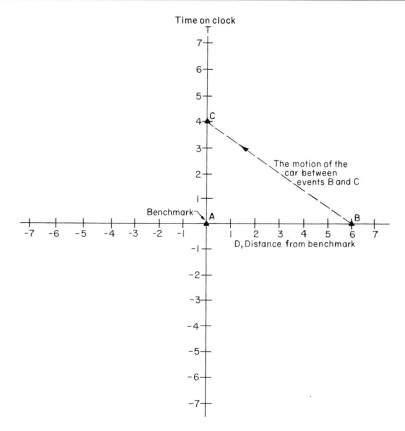

FIGURE A.8

because it moves in space and time, so you at the benchmark and Pablo on the 6-meter marker have paths in spacetime. Even if you both stand still with respect to the tracks, you both progress in time along world lines representing an unchanging position on the track (figure A.11).

Consider next the sequence of events specified in table A.3 (p. 215) and shown in figure A.12.

In this Minkowski diagram, the vertical lines represent time intervals during which the car is not moving. Lines tilted from the vertical indicate motion along the track. Notice that the lines illustrating the car's initial movements toward Pablo (event B to event C) and away from Pablo (event D to event E) are tilted much more with respect to the vertical than is the

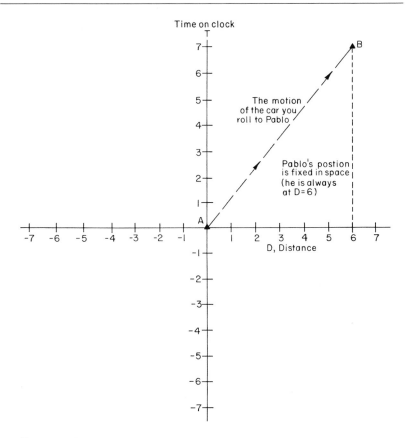

line describing the car's slow roll (after event F). In Minkowski diagrams the tilt of a world line indicates the speed of travel of the object represented by the line. In figure A.13 we show lines representing cars rolling with two different speeds along the track toward Pablo. The greater the tilt of the line from the vertical, the more distance is covered by the car in a given time interval or, in other words, the greater the speed of the car.

Our discussion of Einstein's 1905 relativity paper demonstrated that light and the speed of light play a crucial role in relativity theory. It is therefore important that representations of light in Minkowski diagrams be clearly understood. In such diagrams it is customary to adjust the scales of the time and distance axes so that the path followed by a ray of light is a line

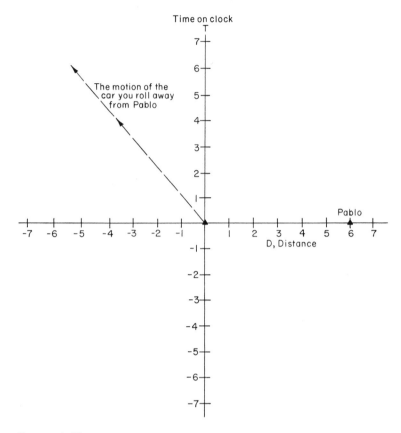

FIGURE A.10

tilted at 45 degrees to these lines.[1] Figure A.14 shows the motion of rays of light sent out by you at the benchmark toward and away from Pablo.

[1] The time and distance scales used in all of the diagrams of this book are arbitrary. Some readers might wish to use our diagrams to determine the sizes of various effects in actual units such as meters and seconds. The diagrams have been drawn with care so that this is in fact possible. However, the condition that the world lines of light be inclined at 45 degrees to the time (or distance) axes establishes a definite relation between the time and distance scales, and this relation must be taken into account if actual units are to be assigned to the diagrams. One can choose any distance units and assign those directly to the distance axis. The unit markings on the time axis will then be given by the following formula: UNITS ON TIME AXIS = (ACTUAL DISTANCE UNITS CHOSEN)/(SPEED OF LIGHT IN CHOSEN TIME AND DIS-TANCE UNITS). For example, suppose that one wants to use the diagrams to solve problems involving distances measured in meters and time in seconds. Then the distance axis can be

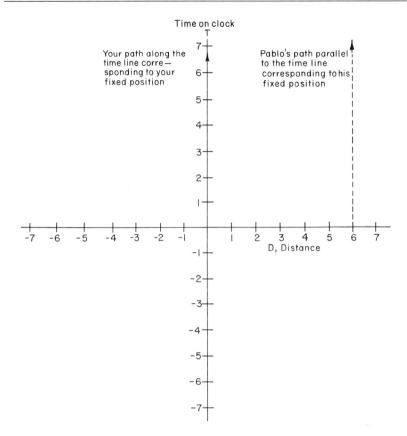

Figure A.11

Let us now use this representation of light to illustrate a simple experiment in a Minkowski diagram. This particular diagram will be valuable later in this appendix when we contrast *measurements* that you make of events with what you actually *see*, and in the next two appendixes as we prove certain assertions made earlier in the book. The experiment consists of the sequence of events listed in table A.4 (p. 215). Figure A.15 shows a Minkowski diagram of this experiment. Again, notice that the lines representing light beams are always tilted at 45 degrees to the vertical.

read directly in meters and the time axis will be read in units, DISTANCE UNIT/SPEED OF LIGHT = 10^{-8} SECONDS, where we have used the fact that in the real units we have chosen (meters and seconds) the speed of light is 10^8 METERS/SEC.

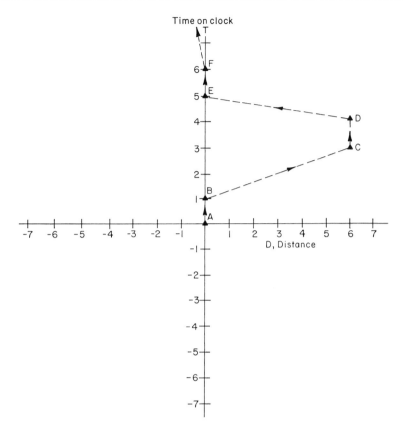

FIGURE A.12

A.3 THE LORENTZ TRANSFORMATIONS

To this point we have been describing things as they would appear to you, at rest on the track. But suppose that Gertrude is again riding a very fast railway car. We know from chapter 3 that she will not measure things as you do and that the Lorentz transformation equations tell us how her measurements of the world differ from yours. In this section we will demonstrate the use of Minkowski diagrams to determine Gertrude's measurements of length and time without the equations of the Lorentz transformations.

We begin with the basic Minkowski diagram that we have used before to

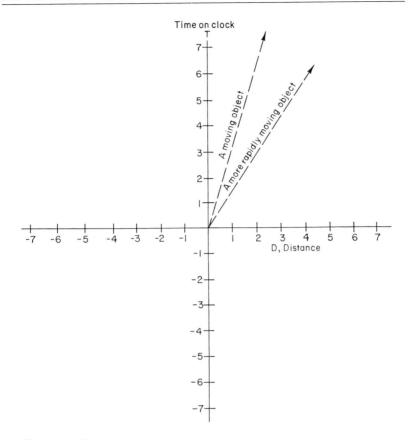

FIGURE A.13

represent events on the railroad track. In figure A.16 we draw the distance
(D) and time (T) lines that serve to locate when and where you measure
events to take place. In what follows, you are at rest on the benchmark as
usual, and Gertrude moves uniformly with respect to you and happens to
pass your location at exactly time 0 on your clock and on hers.

Given these conditions, we can now add a second set of time and distance lines to the diagram (figure A.17). These new lines will be used to
indicate Gertrude's measurements of where and when events take place.
We cannot explain *why* these lines represent Gertrude's measurements
without some algebra; appendix B gives a full mathematical justification
for the following discussion.

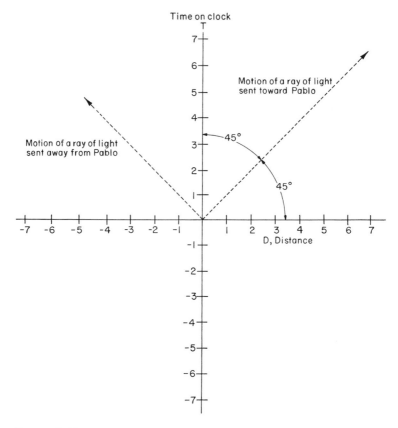

FIGURE A.14

In figure A.17 we have called Gertrude's time measurements T′ and her distance measurements D′ to distinguish them from yours. The amount of tilt in these lines depends on Gertrude's speed: the faster she moves, the greater the tilt. We also point out that the angle between the T- and T′-lines is equal to the angle between the D- and D′-lines in these diagrams (this fact is explained in appendix B). The lines drawn in figure A.17 represent Gertrude moving past you at half the speed of light.

We may use Gertrude's lines drawn in this way to reflect events from her point of view, just as readings you make at rest on the track are correctly reflected in the horizontal distance line and the vertical time line. In other words, the tilted time and distance lines represent the essence of the Lo-

TABLE A.3

Event	Common Time	Position	Description
A	0	0	A car stands still on the benchmark
B	1	0	Car starts to roll quickly to Pablo
C	3	6	Car reaches Pablo and stands still
D	4	6	Car starts to roll away from Pablo
E	5	0	Car reaches the benchmark and is still
F	6	0	Car rolls slowly away from Pablo

rentz transformations in graphical form. There is another detail to mention. The scales marked on Gertrude's time and distance lines are not the same as on yours; that is, the spacing between the distance and time markers along the T'- and D'-lines is not the same as along the T- and D-lines. In fact, the spacing is larger on Gertrude's lines, with the difference becoming more pronounced as her speed increases.

TABLE A.4

Event	Common Time	Position	Description
A	0	0	You send a flashlight beam to Pablo
B	6	6	The light ray hits a mirror that Pablo is holding and the light ray is reflected back to you
C	12	0	The reflected light ray hits you

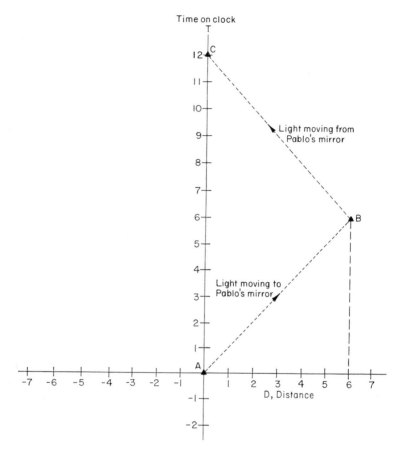

FIGURE A.15

Although the explanation for this requires algebra, we can make it apprehensible in the following nonalgebraic way with the help of figure A.18. At very low speeds we expect that you and Gertrude will differ very little in what you measure. Therefore, at low speeds, her lines are very close to yours (the tilt is small), and the spacing between the time and distance marks on the two sets of lines is nearly the same. As Gertrude's speed increases, we know from relativity theory that your measurements of events and hers will differ increasingly. Therefore, as Gertrude's speed increases, her time and distance lines tilt more away from yours, and the spacing between her and your time and distance marks differs more and more.

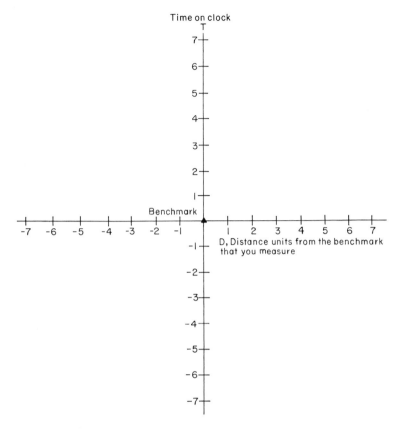

FIGURE A.16

Let us now see just how to read distance and time measurements from Gertrude's tilted lines. Suppose you determine that an event that we call A took place at 5 distance units and 4 time units. You locate this event in the Minkowski diagram shown in figure A.19. To read the time and distance measurements you make for the event you follow the two rules shown in box A.1. The lines specified by these two rules are shown in figure A.19 as the dashed vertical (corresponding to rule 1) and horizontal (corresponding to rule 2) lines.

How would this same event look to Gertrude in her speeding car? In figure A.20 we first draw the tilted time and distance lines discussed above. Here we will assume that she moves past you at half the speed of light. We

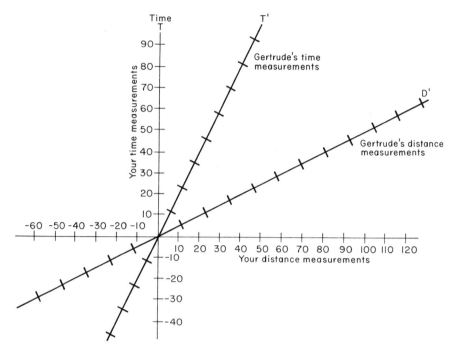

FIGURE A.17

Box A.1
RULES FOR SOMEONE AT REST ON THE TRACK

RULE 1. Draw a line through the *point representing the event*, parallel to the time line. The point at which this line crosses the distance line gives your distance measurement for the event.

RULE 2. Draw a line through the *point representing the event*, parallel to the distance line. The point at which this line crosses the time line gives your time measurement for the event.

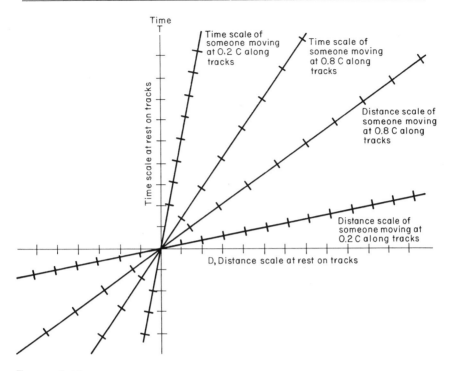

FIGURE A.18

also assume that you and Gertrude pass each other at time 0 on your (and her) clock. The dashed lines used in figure A.19 to locate the time and distance values that *you* measured for event A are also shown in figure A.20; but note that the intersections of these dashed lines with Gertrude's time (T′) and distance (D′) lines are not significant. Only intersections of these dashed lines with *your* time (T) and distance (D) lines have significance.

The rules for finding the time and distance of an event as determined by Gertrude may be expressed in words almost identical to those given in box A.1 that were used to find your measurements of the same event. These new rules are stated in box A.2. The lines specified by these two rules are shown as dashed lines in figure A.21 (the dashed lines corresponding to the application of the rules you used to locate the event no longer appear). The points at which these two dashed lines intersect Gertrude's time and distance lines give her time and distance measurements. Note that if the dashed lines in figure A.21 were extended, they would eventually intersect *your* time and distance lines, but these intersections would have no signif-

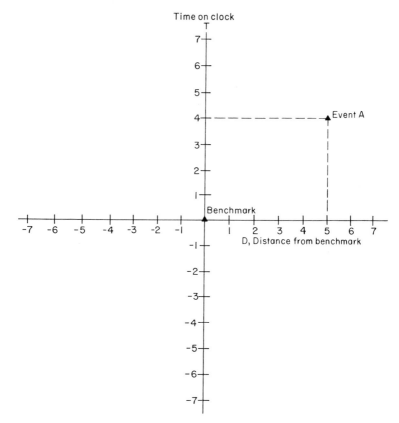

FIGURE A.19

Box A.2
RULES FOR SOMEONE MOVING ALONG THE TRACK

RULE I. Draw a line through the *point representing the event*, parallel to Gertrude's distance line. The point at which this line crosses Gertrude's time line will be her time measurement.

RULE II. Draw a line through the *point representing the event*, parallel to Gertrude's time line. The point at which this line crosses Gertrude's distance line will be her distance measurement.

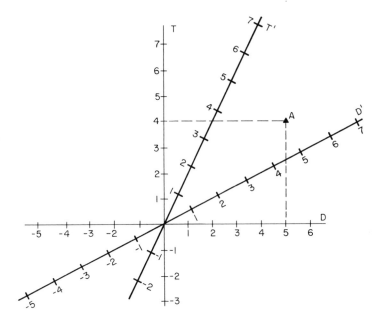

FIGURE A.20

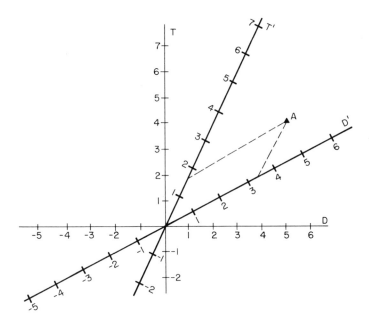

FIGURE A.21

icance. It is only the intersection of each of these tilted dashed lines with its corresponding tilted axis that has significance, and so we have stopped each dashed line at its only significant point of intersection. These intersection points show that Gertrude determines event A to occur at about 1.7 time units and at a distance of about 3.5 units from the benchmark. These are the results we would have obtained by calculations using the Lorentz transformation equations.

Notice in the last two figures that at the benchmark you and Gertrude show the same time value (0) and the same position (0). This meeting between you and Gertrude is called event O in figure A.22. In other words, Gertrude passes you at this instant so that your positions are coincident and both your clocks read zero.

At a later time on your clock, Gertrude will have moved some distance beyond the benchmark. For example, from the speed corresponding to the tilt of Gertrude's lines in figure A.22 we can calculate what her location will be according to you at time 4 on your clock. This turns out to be D = 2 units (at event B) shown in figure A.23.

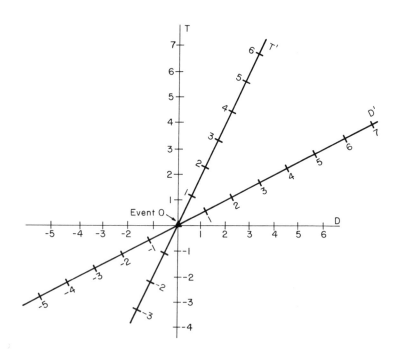

FIGURE A.22

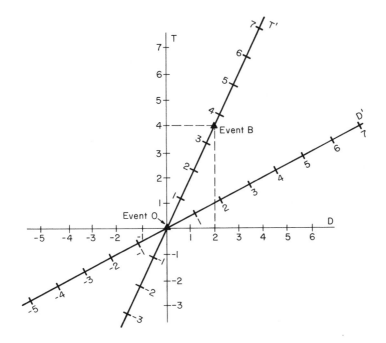

FIGURE A.23

We have again drawn dashed lines showing *your* determination of the time and distance for event B (following rules 1 and 2 in box A.1 discussed previously). It bears repeating that only the intersection of these dashed lines and *your* time and distance lines has significance. In particular, the intersection of the vertical dashed line and Gertrude's distance (D') axis is of no significance. Notice that both events O and B representing Gertrude's position fall on Gertrude's time line. Why is this? If we follow rule II in box A.2 telling us how to determine the position at which Gertrude measures events, then any and all events falling on Gertrude's time line take place at D' = 0, that is, at Gertrude's location. Thus, Gertrude's time line (the T'-line) shows her location at any time, and it is what we called her "world line" in chapter 3. This fact defines the procedure for drawing the T'- and D'-lines in the Minkowski diagram. Knowing Gertrude's speed relative to us we can draw her world line (the T'-line). We have also noted that the D'-line falls at the same angle to the D-line as the T'-line does to the T-line. Therefore, knowing Gertrude's speed, the D'- and T'-lines may be drawn on the Minkowski diagram; these lines may then be used to solve graphically the Lorentz transformations.

A.4 USING THE MINKOWSKI DIAGRAM TO PREDICT RELATIVITY EFFECTS

Now that we have stated the rules for determining Gertrude's distance and time measurements from the Minkowski diagram, let's use them to predict some of the consequences of relativity theory that we discussed in chapters 3 and 4.

Time Dilation. First, recall that you will measure a clock's rate to be slower as it moves past you more rapidly. Suppose you stand on the benchmark with a clock. We will call two consecutive ticks of the clock events A and B; both take place at the same point in space according to you ($D = 0$), and we suppose that you measure the ticks to be exactly one time unit apart. Let us see how Gertrude would view this same clock by locating events A and B in the Minkowski diagram shown in figure A.24.

We follow rules I and II of box A.2 for determining Gertrude's measurements of these two events. In figure A.24 we have drawn the lines specified by these rules as dashed lines (again, remember that only intersections of these lines with *Gertrude's* time and distance lines will be of significance in the figure). She measures event A to happen at the same time and place as you do, because she is assumed to be passing by you at the time event A happens. But from the intersections of the dashed lines with Gertrude's time and distance lines we find that event B, according to Gertrude, happens some time after she has passed you by and at some 0.58 distance units *behind* her on the track (the minus sign on the D'-value indicates that the event takes place behind her). Gertrude's distance determination makes sense since Gertrude leaves you and the clock behind her after event A. What makes less sense to someone unfamiliar with relativity is the fact that Gertrude measures event B to happen at a time $T' = 1.15$ time units. In other words, Gertrude determines that your clock ticks once each 1.15 time units, not each 1.0 time units as you determine. She therefore concludes that your clock is slow.

Now let's try a similar experiment, this time focusing attention on Gertrude's clock instead of yours. We label two successive ticks of Gertrude's clock riding on her car as events C and D, and again we locate these in a Minkowski diagram (figure A.25). Event C takes place just as you and Gertrude pass by one another, and so you both see it happen at the same time and location. She determines that the next tick, event D, happens at the same distance, for she is at rest with respect to the clock, but you measure it to happen after Gertrude and the clock have moved past you 0.58 distance

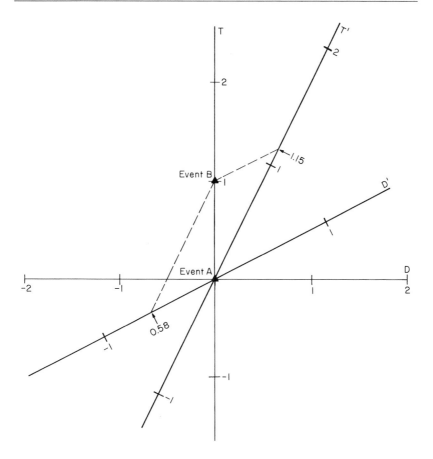

FIGURE A.24

units. You also determine that the next tick happens at a time of 1.15 time units. You therefore conclude that Gertrude's clock is slow.

True to Einstein's principle of relativity for inertial observers, it makes no difference whether you or Gertrude holds the clock under study. Your and Gertrude's respective reference frames are fully equivalent. Each of you will conclude that the other's (moving) clock ticks once each 1.15 time units while your own (at rest) ticks once each time unit.

The clock runs slow according to any observer moving past the clock; the clock ticks accurately to anyone at rest with respect to the clock.

Simultaneous Events. Next consider the observation of simultaneous

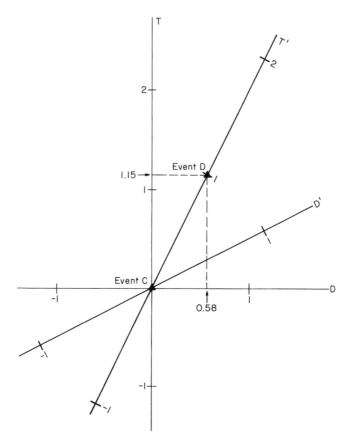

FIGURE A.25

events. You and Pablo are at rest on the track and you both determine that two events, A and B (specified in table A.5), are simultaneous. We represent these events on a Minkowski diagram in figure A.26.

What does Gertrude conclude about the time and location of these same two events? In figure A.27 we add to the Minkowski diagram the lines representing Gertrude's time and distance scales and follow rules I and II of box A.2. Gertrude sees A happen at time 0 on her clock and at the benchmark (she must, because you, she, and event A are all at the same spot); but she concludes that event B happens some 3.6 time units *before* event A. So to Gertrude the two events are not simultaneous. As noted in chapter 3, events that are simultaneous to one observer will not be simultaneous to another observer moving with respect to the first.

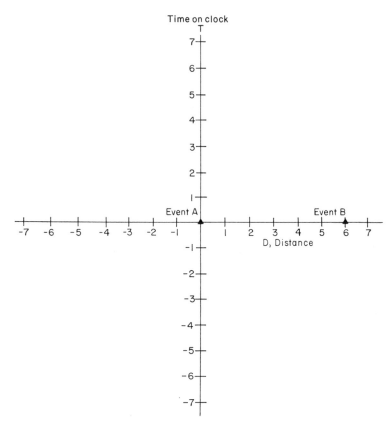

The Temporal Order of Events and Causal Connections. In section 4.8 we stated without proof the conditions under which two events will follow the same temporal order according to the measurements of any and all inertial observers. It is possible for such events to be related causally. We are

TABLE A.5

Event	Common Time	Position	Description
A	0	0	You fire a flashbulb
B	0	6	Pablo fires a flashbulb

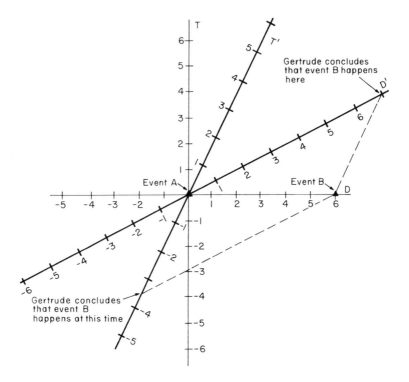

FIGURE A.27

now in a position to offer proof of these conditions that permit causal connections between events. We caution readers that this discussion and its associated figures are somewhat more involved than those in the previous sections of this appendix.

In figure A.28 we show an event A, which, according to you, takes place 8 time units and 4 distance units after event O; event O, in turn, represents Gertrude passing your position on the track. We also show in this figure the results of three experiments involving you and Gertrude: in experiment I Gertrude moves by you at 20 percent of the speed of light; in experiment II she moves by you more rapidly, at 50 percent of the speed of light; and finally, in experiment III she moves by you very rapidly, at 80 percent of the speed of light. In each case Gertrude determines that event A takes place after event O. This will be true for Gertrude no matter how fast she moves. In experiments I and II Gertrude concludes that event A occurs a shorter time after O than according to you; in experiment III it happens at the same

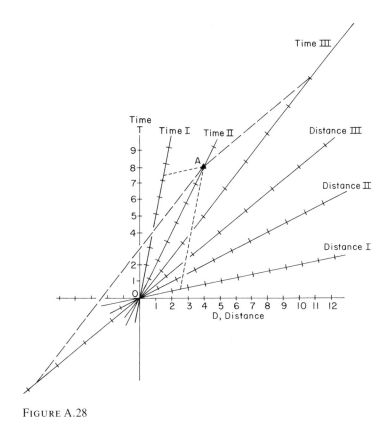

FIGURE A.28

time for Gertrude as for you.[2] But event A always happens after event O. Notice also that when Gertrude moves at the speed corresponding to exper-

---

[2] This may not seem correct. Both you and Gertrude measure the same time for events O and A in experiment III. But as we have shown (section 3.7) your two clocks cannot tick at the same rate. In fact, figures A.24 and A.25 in this appendix used Minkowski diagrams to confirm this fact. There is a significant difference, however, between the situation sketched in figure A.28 and that of the clocks in figures A.24 and A.25. In the latter cases the events refer to two successive ticks of the same clock located at rest *at the same point in space* in either your reference frame (figure A.24) or in Gertrude's reference frame (figure A.25). The two of you disagree on the times of the two successive ticks (events), as our discussion in section 3.7 showed that you must. In figure A.28 we are dealing with two events (O and A) taking place *at different points in space* according to both you and Gertrude. It is possible under such circumstances for Gertrude to move at just the right speed and in the correct direction to yield a time measurement for these two events equivalent to yours (she will disagree with you about where in space the events take place). But this equality of time measurements will *only* be true for two events separated by that unique distance in space. Time

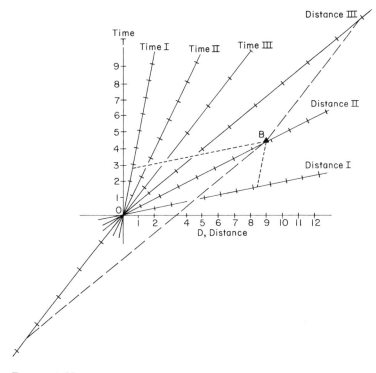

FIGURE A.29

iment II, she determines that event A occurs exactly at the benchmark, whereas you determine that it takes place 4 distance units beyond the benchmark. In experiment I she concludes that event A takes place beyond the benchmark, but closer to it than you determine; and in experiment III, Gertrude's measurements show that the event happens at a point on the track before the benchmark. For an event like A according to Gertrude, the temporal order is fixed with respect to event O, but the spatial location can be either before or after or coincident with that of O.

Contrast this situation with that shown in figure A.29. We locate an event B at $D = 9$ and $T = 4.5$ and carry out the same three experiments with Gertrude moving past your location at 20, 50, and 80 percent of the speed of light. In experiment II event B is simultaneous with O according to Ger-

measurements of other pairs of events (for example, successive ticks of any clock at rest with respect to you or Gertrude) will not result in the same time measurements by you and Gertrude.

trude. In experiment I event B occurs after event O, and in experiment III event B occurs before event O. On the other hand, the spatial relation between events B and O is always the same: event B is always beyond the distance of event O. There is a symmetry between this situation and that sketched in figure A.28. In figure A.28 the event A was fixed in its temporal relation with event O but could assume any spatial order depending on Gertrude's speed. In figure A.29 the spatial order of event B with respect to event O is fixed, but the temporal order of event B may be before or after or simultaneous with event O, depending on Gertrude's speed.

Clearly, were there some causal connection between events O and B in figure A.29, Gertrude could cause havoc. She could determine that a cause (event O) either precedes or follows its effect (event B); she could even determine that the cause and the effect are simultaneous. Fortunately, Gertrude is prevented by the speed of light from committing this outrage. Remember that the world lines for light flashes are represented by lines inclined to your vertical time line in a Minkowski diagram at an angle of 45 degrees. Such a line representing a light signal sent out from event O is sketched as a dashed line in figure A.30, which also shows events A, B, and O. In order for event O to cause another event, some sort of signal must leave O and travel to the location (in spacetime) of the second, related event. We could draw a second dashed line in figure A.30 to represent this signal: it would be a straight line from event O to the related event. Suppose we connect events A and O by such a line. The line is tilted with respect to your vertical time line by less than 45 degrees, meaning that it represents a world line of something moving more slowly than (its world line has less tilt than that of) a light ray. Such a world line between event O and event A represents a possible signal because it corresponds to a signal speed less than that of light. But a line from O to B is inclined at a much greater angle than 45 degrees to your vertical time line, corresponding to a world line of something moving with a speed greater than that of light.[3] As long as light

---

[3] It is essential to bear in mind that this discussion concerns world lines of *signals* in the Minkowski diagram. Any such line tilted at more than 45 degrees to your vertical time line corresponds to something moving with respect to you at a speed greater than that of light. On the other hand, Gertrude's *distance* lines (for example, the three shown in figure A.29) can appear in the same region of the Minkowski diagram that corresponds to world lines of signals moving at speeds greater than light. But the world line of something and Gertrude's distance line are two very different sorts of lines. Gertrude's distance lines are mathematical constructions used to solve the Lorentz transformations without using algebra; hypothetical world lines of signals (or of objects) moving at various speeds may fall in the same region of the Minkowski diagram as Gertrude's distance lines, but they are unrelated to the distance

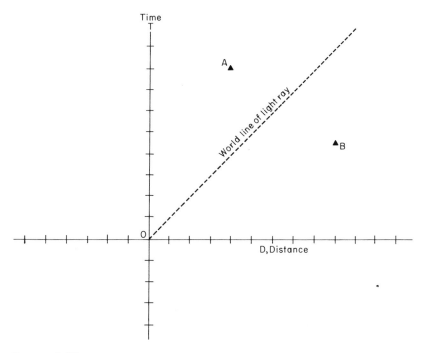

FIGURE A.30

speed remains an ultimate upper speed limit, there can be no possible causal relation between events O and B, and the fact that Gertrude determines these two events to occur in any temporal relationship can have no bearing on causality. On the other hand, it is absolutely necessary that the temporal relation between events O and A be fixed for Gertrude and all other observers, since such events can be causally related by signals moving at speeds less than that of light.

Let us summarize this discussion with another Minkowski diagram. It is clear that events located anywhere in the unshaded portion of figure A.31 can be reached by a signal moving at or below the speed of light from event O, and so these events can be causally related to event O. Similarly, the Lorentz transformations demonstrate that it is just in this region of the Minkowski diagram that events will maintain their fixed temporal sequence to

---

lines. Gertrude's *time line*, however, is also her world line and represents her motion. Because of the speed limit represented by the speed of light, her time line, like any world line, can never be tilted more than 45 degrees with respect to your vertical time line (this is consistent with her three time—or world—lines in figure A.29).

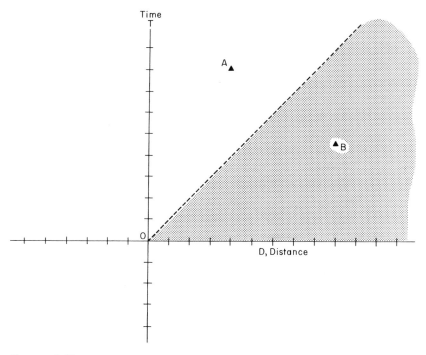

FIGURE A.31

all inertial observers. On the other hand, the Lorentz transformations predict that events that are located in the shaded region of the Minkowski diagram shown in figure A.31 can happen in any temporal order with respect to event O (the order will depend on the speed of the inertial observer). In this same region, any signal traveling between O and another event (represented by a straight world line between O and that event) would have to move at a speed greater than that of light (because the tilt of its world line is greater than 45 degrees to the vertical time axis), so that a causal connection is impossible.[4]

We discuss events in the shaded region of figure A.31 in appendix C, where we speculate on the consequences of things moving faster than light.

What Observers Actually See. Return to the situation depicted in figure

---

[4] Events can occur, of course, at any point in the Minkowski diagram depending on your measurements of distance and time. The shaded region of figure A.31 in no way represents a ''forbidden region'' for *events*. Any event that is *causally connected to event O*, however, cannot fall in the shaded region. Such an event is too distant from event O for a signal to reach it in time for a causal connection.

A.26. You and Pablo report events A and B as simultaneous, but you don't actually see the events at the same instant because it takes light a finite amount of time to reach you from Pablo's position; similarly, it takes a finite amount of time for the light from your flashbulb to reach Pablo. So you actually *see* your event first, and then, later, you see the light from Pablo's. Nevertheless, the two of you know that the events are simultaneous for two reasons: First, your watches are synchronized to the common time system used by everyone at rest along the tracks, and, by prior agreement, you both fired your bulbs at common time zero. Second, you can measure the interval of time between your flash (event A) and your reception of Pablo's flash; since you know the distance separating you and Pablo (6 distance units) and the speed of light, you can calculate when the flash must have left Pablo's position. Your calculation would lead you to conclude from your observations that the two events are simultaneous.

You would locate both events in the Minkowski diagram as shown in figure A.26. We can use the Minkowski diagram to determine what you would actually see by drawing in lines that represent light rays moving from any event to your location (or to Pablo's or Gertrude's to determine what they would actually see). For example, in the case just mentioned, we could draw the world line for light leaving event B and striking your eye as shown in figure A.32. We see that the light from event B reaches you 6 time units after event A. But, to repeat, although the flashes from A and B reach you at different times, you are still able to conclude that they are simultaneous in that they both happen at the same common time.

In general, to determine when things are actually seen by observers (that is, when signals reach them), one need only draw the appropriate world lines for light rays in the Minkowski diagram showing the events; in our diagrams these lines are inclined at 45 degrees to the T- or D-lines.

A.5 THE INVARIANT INTERVAL AND MINKOWSKI DIAGRAMS

In section 4.3 we mentioned that despite the fact that you and Gertrude will disagree on where and when events occur, there are some things on which you will agree. Both of you will agree on the value of the speed of light measured in your respective reference frames. And both of you will agree on a number called the *interval* separating any two events in spacetime (or, equivalently, in the Minkowski diagram).

One may think of this interval as a measurement of separation of events

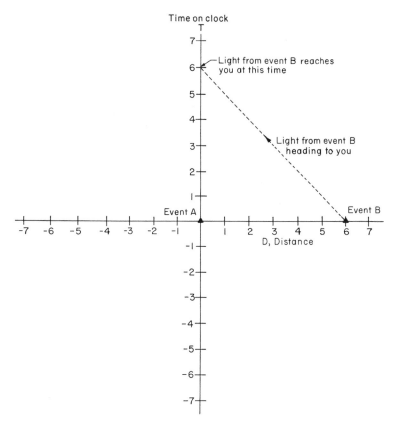

FIGURE A.32

in four-dimensional spacetime. The formula for the interval between any two events is given by the equation

$$(\text{INTERVAL})^2 = (\text{SPATIAL SEPARATION OF EVENTS})^2 -$$
$$(\text{TEMPORAL SEPARATION OF EVENTS})^2.$$

For example, consider the two events A and B in figure A.24. You see a spatial separation of zero (both events happen at the same place according to your measurements) and a temporal separation of 1.

Gertrude measures a spatial separation of -0.57735 (a number rounded off in figure A.24 to -0.58) and a temporal separation of 1.15470 (rounded off to 1.15 in the figure). Substituting these separations into the equation defining the interval, you calculate that

$$(\text{INTERVAL})^2 = 0 - 1 = -1,$$

and Gertrude calculates

$$(\text{INTERVAL})^2 = 0.33333 - 1.33333 = -1.$$

You and Gertrude measure different places and times for the events, but the interval between them is the same for you both (it is an invariant).[5]

THIS ends our discussion of the Minkowski diagram and its use in evaluating the Lorentz transformations. Anyone interested in the proof of rules I and II of box A.2 should read the next appendix. Readers interested in the consequences of things that are able to go faster than light (tachyons) should read appendix C (which does *not* depend on the discussion in appendix B).

---

[5] Notice that in calculating the interval it is important not to round off the measured time and distance values excessively, or the calculated intervals will not agree exactly.

Appendix B

Explanation of Rules I and II for Solving the Lorentz Transformations in Minkowski Diagrams

In appendix A we discussed the graphical representation of events by locating T and D on the perpendicular time and distance lines of the Minkowski diagram, and we stated that by following the two rules given in box A.2, a solution to the Lorentz transformation equations could be achieved by graphical means. We will now prove the validity of these rules by algebraic argument. What follows here requires a knowledge of algebra, but the manipulations are not advanced.

Stated in the language of algebra, the essence of the Lorentz transformations is this: We are given measurements T and D of time and distance for an event made by one inertial observer (for example, you at rest on the tracks); we wish to know the values of these entities measured by another inertial observer moving with respect to the first (for example, Gertrude). This problem is solved using the equations of the Lorentz transformations. In these equations we have called the measurements made by the moving observer T' and D'

$$T' = \frac{T - VD/C^2}{\sqrt{I - V^2/C^2}}$$

$$D' = \frac{D - VT}{\sqrt{I - V^2/C^2}},$$

where V is the relative speed of the two observers and C is the speed of light. Given any T and D we can find T' and D' by simple substitution.

In what follows, it will be important to notice that all events represented by points falling on the T-line in the Minkowski diagram are events that happen at D = 0, that is, they all happen at the same point in space. Similarly, all events happening at a distance D $=$ 1 fall along a line parallel to the T-line but displaced from it by one distance unit, and so on for events happening at other distances D $=$ 2, 3, 4, 5, . . . (see figure B.1)

All events represented by points falling on the D-line happen at time T $=$ 0, that is, they are simultaneous. All events happening at T $=$ 1 are also simultaneous; they all fall along a line parallel to the distance line but displaced from it by one time unit; and so on for events happening at other times T $=$ 2, 3, 4, . . . (see figure B.2).

As we discussed in appendix A, the Minkowski diagram can also be used to represent the motion of objects. Suppose, for example, that you carefully observe the motion of a point on Gertrude's car as it moves down the track. If this point passes your benchmark (D $=$ 0) at T $=$ 0, you can write the following equation for the distance of Gertrude's car at any time:

$$D = VT.$$

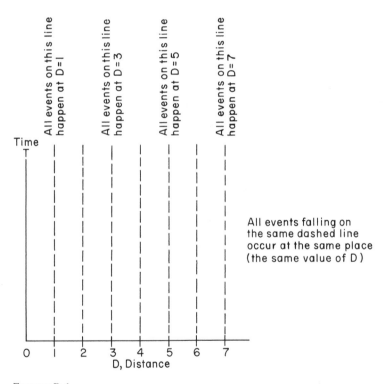

FIGURE B.1

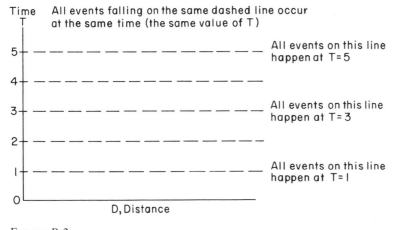

FIGURE B.2

This equation can be represented as a line on the Minkowski diagram (figure B.3). This line shows how the distance of Gertrude's car from your benchmark (at $D = 0$) increases as time (T) goes on. If V is increased, Gertrude will cover a greater distance in any time T, and so the line representing her travel will have a greater tilt away from the vertical time line (or, in algebraic jargon, since $D = VT$, the greater the value of V, the greater the *slope* of the line).

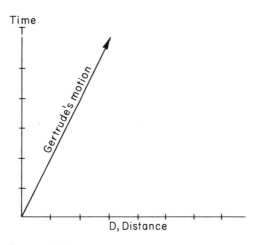

FIGURE B.3

We mentioned in a footnote to appendix A that the scales of the D- and
T-lines in our Minkowski diagrams have a special relationship to one another. Most of the Minkowski diagrams used in this book are drawn with the
D- and T-scales so that the speed of light has the value unity ($c = 1$). With
this choice, lines representing the path followed by light rays will always
be inclined at an angle of 45 degrees to the D- or T-lines.

Suppose that we locate an event on a Minkowski diagram (figure B.4)
from the D- and T-values that you measure for it. As we will now see, it is
possible to draw lines on this diagram corresponding to Gertrude's distance
and time scales, and to use these to determine how any given event will be
measured by Gertrude.

Gertrude's distance line and yours should be analogous. We can express
this analogy mathematically by recalling that your D-line may be defined as
that along which all events happen at $T = 0$ according to you. How do we
express this definition in terms of Gertrude's measurements? We know that
in general,

$$T' = \frac{T - vD/c^2}{\sqrt{1 - v^2/c^2}} \,.$$

To turn this equation into a definition of Gertrude's distance line (the D'-
line), we will set $T' = 0$, in analogy to the algebraic relation that defines
your D-line. Since $c = 1$, our defining equation becomes,

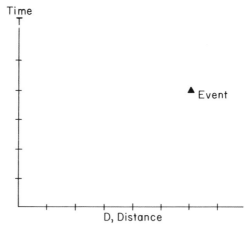

Time
T

▲ Event

D, Distance

FIGURE B.4

$$0 = \frac{T - VD}{\sqrt{1 - V^2}} \quad \text{or,}$$

$$T = VD$$

which is an equation representing a straight line on the Minkowski diagram. This line, labeled D′, is shown in figure B.5 where we have assumed some arbitrary value for V.

Similarly, for events occurring at other times, according to Gertrude (for example, T′ = 1, 2, 3, etc.), we can write,

$$\text{CONSTANT} = T′ = \frac{T - VD}{\sqrt{1 - V^2}} \quad \text{and so,}$$

$$(\text{CONSTANT}) \sqrt{(1 - V^2)} = T - VD$$

$$(\text{CONSTANT}) \sqrt{(1 - V^2)} + VD = T.$$

This last equation represents a series of straight lines all having the same tilt as the D′-line (that is, all parallel to the D′-line) but displaced from it vertically by an amount,

$$(\text{CONSTANT}) \sqrt{(1 - V^2)},$$

as shown in figure B.6. All events falling on the same dashed line occur at the same time according to Gertrude.

Notice that the event that took place at rest with respect to you on the

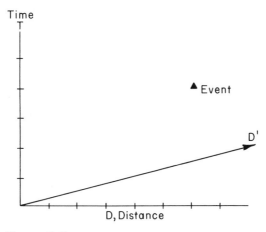

FIGURE B.5

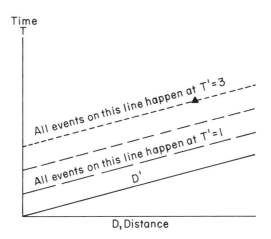

FIGURE B.6

tracks (and previously identified in figures B.5 and B.6) intersects the line corresponding to $T' = 3$. This, then, is the time at which Gertrude reports the event as having taken place.

Refer to rule I of box A.2. We can now see that by drawing a line through the point representing an event parallel to the D'-line, we are really drawing a line along which, according to Gertrude, all events are simultaneous with our given event. But before we can complete our justification of rule I, we must define the time line for Gertrude, the T'-line, since it is the scale on this line, according to rule I, that we use to find out what time value Gertrude measures for the event.

Recall that the line representing the times at which you see events may be defined as the line on which all events take place at $D = 0$ according to your measurements. Other lines corresponding to events at other distances from you are parallel to this line (see figure B.1). Again, we will define Gertrude's time line by mathematically stating an analogy to the definition of your time line; for Gertrude the Lorentz transformations state that

$$D' = \frac{D - VT}{\sqrt{1 - V^2/C^2}},$$

and to define Gertrude's time line in analogy to the algebraic condition defining your time line, we specify $D' = 0$ (and $C = 1$), so that

$$0 = \frac{D - VT}{\sqrt{1 - v^2}} \quad ,$$

$$D = VT,$$

$$T = (1/v)\,D.$$

Again this is an equation for a straight line, and, in fact, it is identical to that describing Gertrude's motion from your perspective. This line is shown in figure B.7.

Comparing the equation for the D'-line (that is, for events all having the time $T' = 0$), with the equation for the T'-line (corresponding to $D' = 0$), we see that the slopes (tilts) of the two lines are v and $(1/v)$, respectively. This means that the angle between the T-line and the T'-line is identical to the angle between the D-line and the D'-line; this angle increases as v increases. These were assertions made without proof in appendix A.

For events at other distances $D' = 1, 2, 3, \ldots$ according to Gertrude, we can write

$$\text{CONSTANT} = D' = \frac{D - VT}{\sqrt{1 - v^2/c^2}} \quad ,$$

$$(\text{CONSTANT})\sqrt{(1 - v^2)} = D - VT,$$

$$T = D(1/v) - (\text{CONSTANT})\sqrt{(1 - v^2)},$$

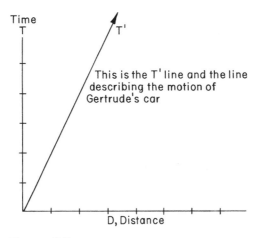

FIGURE B.7

which represents a set of lines having the same slope as the T'-line (that is, they are parallel to it) but displaced from it vertically by an amount equal to

$$(\text{CONSTANT}) \sqrt{(1 - v^2)},$$

as shown in figure B.8. All events falling on the same dashed line occur at the same place according to Gertrude, but not to you.

Notice that our event intersects the line corresponding to D' = 4. This must be the distance at which Gertrude determines the event to take place.

Refer now to rule II of box A.2. We see that by drawing a line through the point representing an event parallel to the T'-line, we are really drawing a line along which all events happen at the same place according to Gertrude. But what is the value for this distance—that is, what is D' for this event?

Notice how each of the dashed lines in figure B.8 intersects the D'-line at different points. These points of intersection may be used to mark the distance scale on the D' line. And now rule II may be completely justified. The point on the D'-scale crossed by a line drawn through an event and parallel to the T'-line must be the D' value for that event.

Similarly, we can use the points on the T'-line intersected by the dashed

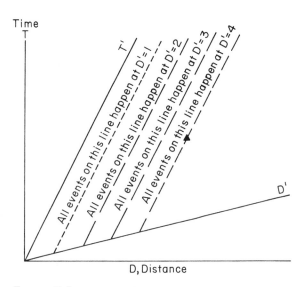

FIGURE B.8

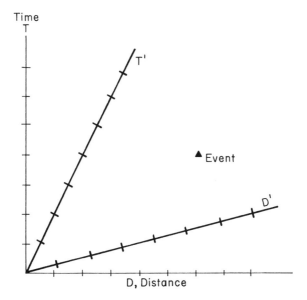

FIGURE B.9

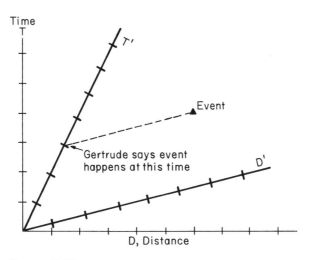

FIGURE B.10

lines representing events taking place at the same time T' to mark off a time
scale on the T'-line. Rule I for finding T' for an event then follows directly.

The full Minkowski diagram showing the event along with Gertrude's
T'- and D'-scales appears in figure B.9. In this figure we have assumed that
Gertrude is moving at half the speed of light.

To summarize, once we know how fast Gertrude is moving, we can draw
lines on the graph representing Gertrude's distance and time scales (the D'-
and T'-lines). By passing a line through a point representing any event parallel to the D'-line, we can read T' from the intersection of the line with the
T'-scale (see figure B.10). By passing another line through the point representing an event parallel to the T'-line, we can read D' from the intersection of this line with the D'-scale (see figure B.11). Thus rules I and II in
box A.2 for solving the Lorentz transformations in the Minkowski diagram
are justified.

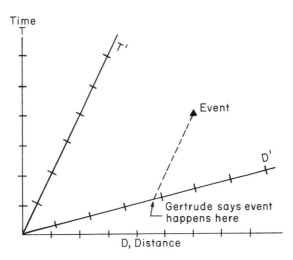

FIGURE B.11

Appendix C

The Trouble with Tachyons

There was a young lady named Bright,
Whose speed was far faster than light
She set out one day
In a relative way,
And returned home the previous night.

—Arthur Henry Reginald Buller

C.1 Introduction

This appendix deals with tachyons, particles that go faster than light.[1] Part of this discussion is highly speculative, and we want to be clear about what is speculation and what is not. No tachyon has ever been detected. Their existence remains unproven. Nevertheless, the *possibility* of tachyons is not ruled out by relativity theory,[2] and the search for these bizarre particles continues precisely because they *might* exist. In this appendix we wish to speculate about some of the consequences if tachyons should one day be discovered.

Figure C.1 shows the report of one experiment designed to search for tachyons.[3] Over the years, physicists have developed a variety of techniques for detecting all sorts of particles and for measuring their speeds. The apparatus used in this particular experiment detected "cosmic rays," particles produced naturally all the time that "shower" down to the surface of the earth. The hope was that sometime, among all the cosmic rays reach-

[1] The word "tachyon" comes from the Greek *tachys* meaning "swift," and was coined in the paper by G. Feinberg, "Possibility of Faster-Than-Light Particles," *Physical Review* 159, no. 5 (1967): 1089–1105. Normal particles moving below the speed of light are sometimes called "tardyons."

[2] This point was discussed in detail in sections 4.4 and 4.5.

[3] D. F. Bartlett, D. Soo, and M. G. White, "Search for Tachyon Monopoles in Cosmic Rays," *Physical Review D* 18 (1978): 2253–2261.

PHYSICAL REVIEW D   VOLUME 18, NUMBER 7   1 OCTOBER 1978

Search for tachyon monopoles in cosmic rays

D. F. Bartlett and D. Soo

Department of Physics and Astrophysics, University of Colorado, Boulder, Colorado 80309

M. G. White

Joseph Henry Laboratories, Princeton University, Princeton, New Jersey 08540
(Received 26 June 1978)

We have searched for a particle which combines the properties of a tachyon with those of a magnetic monopole. The tachyon monopole is assumed to exist in cosmic rays striking the earth and to be influenced by the extensive magnetic fringing field of Fermilab's 15-ft bubble chamber. By hypothesizing that the tachyon monopoles will either emit Cherenkov radiation in air or ionize Lexan plastic we set an upper limit of 5×10^{-12} cm^{-2} sec^{-1} on their flux.

INTRODUCTION

Despite extensive effort, neither tachyons nor magnetic monopoles have been found.[1] Perhaps this failure is because these particles have generally been sought separately.

This paper describes a search for a particle which combines the properties of a magnetic monopole with those of a particle which travels faster than light. Some time ago a search for such a tachyon monopole (TM) was made near a radioactive source.[2] The present search was motivated by the observation that if free tachyon monopoles exist at all they might be found in high-energy cosmic rays of unspecified origin. In contrast, the prospects for finding a tachyon monopole among the particles manufactured by accelerators may not be so sanguine since nothing in current high-energy theory or experiment even hints at the presence of tachyon monopoles.

When interest in tachyons was revived a decade ago,[3] physicists were content to predict how faster-than-light particles would interact with apparatus here on earth.[4] In this context, Huygens' wavelet theory makes it appear reasonable that an electrically (or magnetically) charged tachyon would emit Cherenkov radiation even in vacuo. Experimentalists sought in vain for such radiation.[5]

Recently, however, several physicists have formulated extended theories of relativity. In these theories there is assumed to be a universe S' of objectives which have velocities less than that of light relative to each other but which have velocities greater than the speed of light relative to our system S. The objects in S' are assumed to obey the normal laws of physics when viewed by an observer in S'. In particular, a light wave emitted in S' propagates isotropically according to the equations

$$ds'^2 = dt'^2 - dx'^2 - dy'^2 - dz'^2 \tag{1a}$$

$$= 0, \tag{1b}$$

where we define $c = \hbar = 1$.

Suppose that, when viewed from our frame, S' is moving with a velocity $v > 1$ along the x axis. How does the world line ds' [Eq. (1a)] of a particle's motion in S' appear in our frame? It is clear that we cannot simply write $ds'^2 = ds^2 = dt^2 - dx^2 - dy^2 - dz^2$ because then a particle which is at rest in S' ($ds'^2 > 0$) would in our frame have $dx^2 < dt^2$ and thus appear to be moving slower than light, contrary to hypothesis. At the very least the signs of the terms in dx^2 and dt^2 must be interchanged. On this there is general agreement, as there is on the specific transformations

$$t = \delta(t' + vx'),$$
$$x = \delta(vt' + x'), \tag{2}$$

where

$$\delta = (v^2 - 1)^{-1/2}.$$

Theorists disagree, however, on the transformations for y and z. If one wishes to have a spherical light wave from a tachyonic source appear spherical to an observer on earth, he must change the signs of y^2 and z^2 to match the change in x^2 and t^2. This procedure yields

$$ds^2 = -dt^2 + dx^2 + dy^2 + dz^2. \tag{3}$$

Such a transformation has been promoted by Ricami and Mignani in a number of articles.[6] Although preserving the invariance of the speed of light, this transformation has the unfortunate consequence that the coordinates normal to the velocity of the tachyon source become imaginary upon transformation,

$$y = iy', \quad z = iz'. \tag{4}$$

Quaternions and exotic numeration schemes have been proposed to give reality to these imaginary coordinates.[7] The theory, however, has as yet been unable to give experimentalists a definitive test.

FIGURE C.1

ing the particle detector on the surface of the earth, some tachyons would be found. The apparatus registered the arrival of particles over a period of time, and if one were found with a speed greater than that of light, a tachyon would have been detected. This experiment failed to detect a single tachyon, as have all other experiments to date, although figure C.2 shows the report of a very "close call."[4] This "detection" was later proven to be an error in interpretation of the data received from the experimental apparatus; but had it been confirmed, it could have plunged physical theory into a state of intellectual havoc.

C.2 Tachyons in the Minkowski Diagram

Now we return to the Minkowski diagram that we have drawn for events taking place along the railroad track to see just how things that move faster than light would be represented. Figure C.3 shows the Minkowski diagram with a dashed world line for a flash of light. Event O marks the transmission of the light flash from the benchmark at time zero on the common system of time used by observers at rest along the track. Objects sent out from the benchmark at the same common time as the light flash, and at any speed less than light will have world lines that pass from the point O to points within the shaded region of the diagram. That is, the tilt of these world lines from the time axis must be less than the tilt of the dashed light line.

A tachyon, on the other hand, sent out from the benchmark at common time zero will be represented by a line that passes from point O to some other point in the un-shaded region of figure C.3. Four such lines are shown in figure C.4, but only two of them are valid representations of tachyons. Lines I and II represent legitimate tachyons. Line III represents something moving with infinite speed: no time is required for such an object to cover any distance. Things moving with infinite speed are *not* what we have in mind when we speak of tachyons. For our purposes, tachyons are things moving faster than light, but with finite speed. On the other hand the sort of object represented by line IV does not move with infinite speed, but there is something else wrong with such a world line. Notice that as the particle moves along the track from the benchmark (that is, as its measured position D increases) time *decreases*—that is, the particle moves "backward" as time goes on. Or, putting it another way, according to world line IV, as

---

[4] Roger W. Clay and Philip C. Crouch, "Possible Observation of Tachyons Associated with Extensive Air Showers," *Nature* 248 (1974): 28–30.

LETTERS TO NATURE

PHYSICAL SCIENCES

Possible observation of tachyons associated with extensive air showers

SEVERAL searches[1-4] have been made for tachyons using either laboratory particle sources or high energy cosmic rays. Effects associated with their supposed characteristic mass and velocity have been searched for but so far no positive evidence has been reported. As has been repeatedly pointed out, however, the goal of these searches is so important that all possible avenues should be fully investigated. We report here apparently positive results from a pilot search for tachyons associated with cosmic ray showers of energy about 2×10^{15} eV.

The first interaction of a primary cosmic-ray nucleon occurs at a typical altitude of 20 km (ref. 5). Further interactions result in a cascade of relativistic particles travelling with speeds close to that of light (c). Thus, most of the particles in this extensive air shower (EAS) arrive at sea level with a time spread of only a few nanoseconds[6]. If any shower particles are produced with velocities greater than c, they should be observable in the time interval up to 20 km/c (60 μs) before the arrival of the shower front. The precise time depends on their velocity and production altitude. A pilot experiment has been conducted to search for particles arriving within this time period. In this experiment, the EAS which were studied had energies some two orders of magnitude higher than in previously reported work[4].

A plastic scintillator was used to detect the particles. It is not clear what interactions a tachyon might have with the atmosphere or the scintillator. The present search was made assuming that tachyons are produced in EAS interactions at heights between 20 km and 400 m (800 kg m^{-2} and 10,000 kg m^{-2}) and have a sufficiently long absorption length for some to reach the detector. Detection might be accomplished by direct interaction of the tachyons with the scintillator or through the production of secondary particles which interact with the detector. It is not necessary that a single tachyon should produce a large response in the scintillator (as, for instance, a charged relativistic particle would). In principle, provided that observations are initiated by the detection of EAS, it is possible to sum results from many observations so that small non-random effects can be observed.

The chief experimental difficulty with this procedure is that if a recording device is triggered by the arrival of an EAS, it will be too late to observe the tachyon unless substantial signal delays are inserted. This was overcome in the present case by the use of a digital transient recorder (Biomation International, Palo Alto, California; model 610B) which enabled us to trigger our recording system from an EAS and then examine the signal from the particle detector recorded prior to the arrival of the trigger. The device continually samples and digitises (to six bit accuracy) the output of the particle detector. Two hundred and fifty-six words of this digitised data are stored in a shift register which is continually updated. Thus, when a trigger is received, the previous 256 words of data are in store (in our case representing 128 μs) and can be output at leisure. Output is to a chart recorder after digital to analogue conversion. In this mode, the recorder only outputs 228 words (114 μs).

The EAS trigger was obtained from the fast timing part of the air shower array at the Buckland Park field station of the University of Adelaide. Five 1 m square plastic scintillators, 50 mm thick, were used, in a square array of side 30 m, one scintillator being at each corner with one at the centre. Each scintillator was viewed with a Philips XP1040 photomultiplier and the arrival of an EAS was detected by a coincidence between the centre detector and any three of the outer detectors with a resolving time of 150 ns. Thus, air showers were detected from a cone about the zenith, with half angle of about 35°. The mean rate of showers was one per 500 s, corresponding to a minimum shower energy of about 2×10^{15} eV. In addition, one of the corner scintillators was also viewed by an RCA 8055 photomultiplier connected to a charge sensitive preamplifier, the output of which was the signal recorded by the transient recorder. The impulse response of the system had a width of 1.7 μs at half maximum ensuring that a sampling interval of 0.5 μs gave an acceptable reconstruction of the waveform on digital to analogue conversion.

Data from a total of 1,307 air showers, detected between February and August, 1973, have been analysed. The aim of the analysis was to demonstrate whether or not non-random effects were observable immediately preceding an EAS and for this reason a simple analysis procedure was employed. The time (with respect to the arrival of each EAS) of the largest excursion of the amplified photomultiplier output was noted. If there was doubt as to which of a number of pulses was the largest, all the apparently equal pulses were included. In practice, approximately 4% of the events had their two largest pulses sufficiently close in amplitude to cause ambiguity. In order to check on observer bias in assigning relative pulse heights, appoximately 600 events were re-read by an independent chart reader. Some 3% of the events had the assignment of the largest pulse changed but no systematic bias was found.

The position of the shower front on the output trace can be adjusted; thus the time interval available for analysis is dependent on the exact setting of the transient recorder. For 1,176 of the events the record extended 105 μs before the air shower arrival. Figure 1a shows a histogram of the positions of the largest pulses for these events. The figure suggests that the distribution of the largest pulses is not uniform and a x^2 test on the data indicates that there is only 1% chance that the data is from a uniform distribution.

In order to include data from all 1,307 recorded events it is necessary to limit the period of analysis to 97.5 μs before the shower arrival. The resulting histogram is shown in Fig. 1b. A x^2 test on this data shows less than 0.1% probability that the data is from a uniform distribution. In addition, since if tachyons are observed one might expect their arrival times to be spread over more than one 7.5 μs bin, one can also test to see whether there are specific time regions contributing excessively to the non uniformity or whether the large x^2 is due to a more or less random time distribution of excesses. This problem has been discussed by David[7] who noted that a calculation of x^2 involved the squaring of deviations from a hypothetical set and that independent information on the sign of the deviations was ignored. An examination of the distribution of bins above and below the mean shows that the probability that our data are selected from a uniform distribution is less than 1%, on the basis of a nonparametric run test.

FIGURE C.2

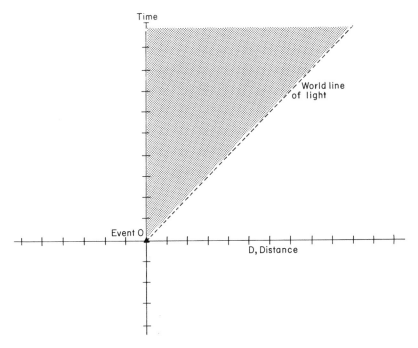

FIGURE C.3

time goes on, the particle seems to approach closer and closer to the benchmark, finally arriving there at time 0 on the common system. This line illustrates a receipt of a particle by you at the benchmark, not your transmission of the particle. So any world lines with tilts from the vertical of 90
degrees or more in the Minkowski diagram are ruled out as possible tachyons. Allowed world lines for tachyons that you send out at common time
zero are, then, lines that pass through point O and fall in the shaded region
of figure C.5.

This section has defined what we mean by tachyons and we have shown
how to represent them in a Minkowski diagram, and so far there does not
seem to be anything particularly interesting about them. But that is because
we have not let Gertrude enter the discussion. Now we allow Gertrude to
move uniformly along the railroad track. Things get very interesting when
we do.

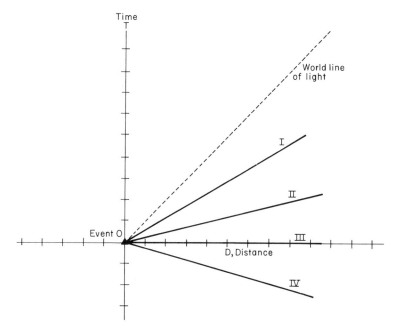

FIGURE C.4

C.3 TACHYONS AND MOVING OBSERVERS

Figure C.6 shows a Minkowski diagram with two events along the world line of a tachyon that you send out along the track. Event O is your act of firing the tachyon. Event A marks its passage past the 9 distance-unit mark on the track. Our discussion in section A.4 (and in particular figure A.29) shows that Gertrude, moving uniformly along the track, can determine that events like O and A take place in any temporal order: O before A, O after A, or O simultaneous with A, depending on her speed. We also mentioned that because communication between events must take place at speeds less than that of light, events like O and A cannot, therefore, be causally related. This seems innocent enough. But now we are supposing that we have something that does move faster than light; let us involve Gertrude directly in the tachyon experiments, as an active participant and not a passive observer.

We are going to reperform the sort of experiment shown in figure A.15 where you and Pablo exchanged flashes of light. But now, instead of exchanging flashes of light with Pablo at rest on the tracks, you will exchange

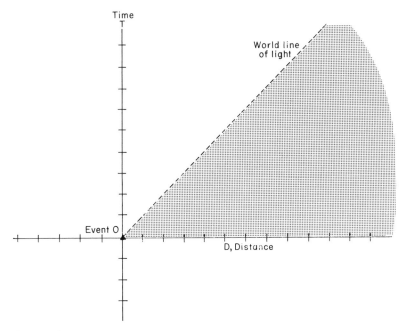

FIGURE C.5

tachyons with Gertrude, who is moving past you at half the speed of light. Suppose that 80 time units after Gertrude passes you on the benchmark, you fire a burst of tachyons down the track chasing after her. We call your firing of the tachyon burst event A and locate it on the Minkowski diagram in figure C.7. The tachyons move at a bit more than seven times the speed of light according to you. The tachyons catch up with Gertrude at event B, 7.2 time units later (again, according to you) and strike her. We further suppose you and Gertrude have a firm agreement that when your tachyons hit her, she is to fire a return volley to you. We assume that Gertrude's response time is sufficiently fast so that she returns the tachyons in your direction just as soon as your tachyons reach her. In other words, event B represents two things: Gertrude's receipt of your tachyons and her firing of the return volley.

Figure C.8 shows the world line of Gertrude's return volley of tachyons. This world line has been sloped to show that according to Gertrude the blast of tachyons travels at nine times the speed of light and strikes you at event C. But there is something strange about this Minkowski diagram. Notice that according to you, event C preceded event B. In other words, as your

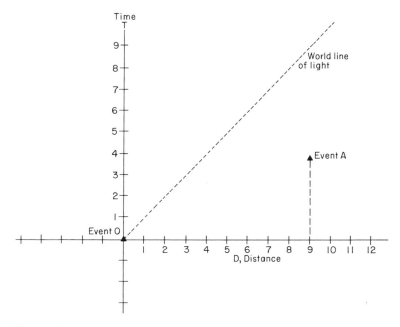

FIGURE C.6

time goes on from event C, the tachyons move away from you and ultimately disappear inside Gertrude's tachyon gun at event B. We have said that Gertrude fires the tachyons to you, but that is not what you determine. According to your measurements, the tachyons start out hitting you and move backwards into Gertrude's gun at slightly less than twice the speed of light! This seems nonsense. Have we drawn an impossible world line for Gertrude's return volley of tachyons? Let's see how things appear to Gertrude.

Figure C.9 shows Gertrude's determination of the time and position of event C, the event marking your receipt of her return volley. The dashed lines are drawn according to the rules in box A.2. From them we see that according to Gertrude, her return volley moves away from her as time goes on at a speed nine times that of light, so that from her perspective this is a blast of perfectly normal tachyons. The fact that you determine these same tachyons to be moving "backwards in time" is of no consequence to Gertrude. She has fired her tachyons in an entirely legitimate way. Your state of motion has made your determination of things differ from hers, but that

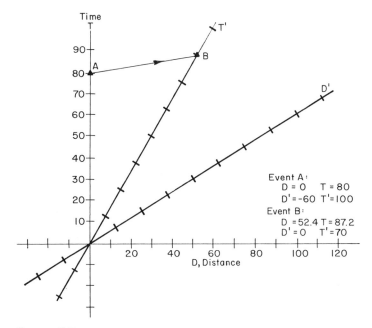

FIGURE C.7

is all, just as we have seen so often in relativity experiments. It is not a question of who is right or who is wrong. The two of you just differ.

To summarize: You have fired a blast of tachyons to Gertrude and, by prior agreement, she returns a burst of tachyons to you. The net result is shown in figure C.10. And now Gertrude has created a problem. According to you, event C (your receipt of Gertrude's return blast of tachyons) has occurred *before* you send out the original blast to Gertrude.

Figure C.11 shows a similar sort of experiment, only this time Gertrude initiates the sequence of events: at event A Gertrude sends a burst of tachyons to you, you receive them at event B, and, by prior agreement, send a return volley of tachyons to Gertrude; she receives these at event C. But event C precedes event A according to Gertrude. This time *she* receives the return burst of tachyons before sending out the original blast to you. Again, in this second example, notice that your return volley is of perfectly ordinary tachyons, moving along toward Gertrude as time goes on at seven times the speed of light. But to Gertrude, these tachyons seem to move "backward" as her time goes on.

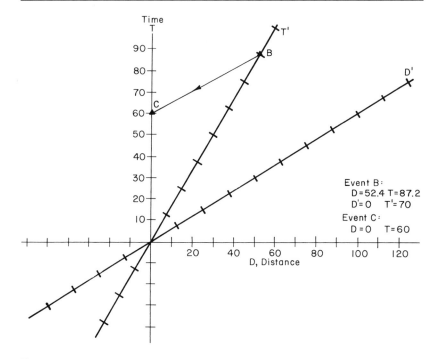

FIGURE C.8

We have encountered other highly counterintuitive results of relativity theory in this book (time dilation in section 3.7 and length contraction in section 3.8), but in everyday life we are able to ignore these phenomena because we do not usually deal with things moving at speeds comparable to that of light where the effects of relativity become noticeable. But with tachyons we are able to create counterintuitive situations that just cannot be ignored. If tachyons are one day found, and if the Lorentz transformations continue to function properly in those situations (and there is no reason to suppose that they would not), violations of causality of the sort just described will not be put aside. Somehow we will have to deal with them intellectually and as a matter of practical necessity.

C.4 What If Tachyons Do Exist?

If tachyons do exist and if experiments of the sort just described can be performed, then we are faced with a strictly illogical situation: effects can

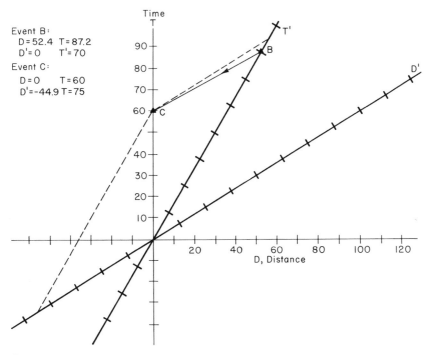

Event B:
D = 52.4 T = 87.2
D'= 0   T'= 70
Event C:
D = 0   T = 60
D'= -44.9 T = 75

FIGURE C.9

precede their causes in some situations. Also notice that in experiencing event C before event A in figure C.10 you are in essence receiving information about the future (your transmission of tachyons later on at event A) before it happens. But suppose that after receiving the tachyons at event C you decide not to send them out to Gertrude at event A later on? In that case you could not have received them at event C in the first place. Does this mean that once you have received Gertrude's tachyons at event C your future is predestined so you must carry out event A? Suppose that event C occurs many years before event A, perhaps even before your birth, then is your life predestined to the time of event A? Would the detection of tachyons usher in a new era of belief in predestination, a belief substantiated by experiments in the physics laboratory? And would this not render the notion of "free will" highly problematic?

In short, with the discovery of tachyons, some of our deepest assumptions about the order of things are threatened or erased. And some have used this very fact to argue that tachyons will never be found. Tachyons

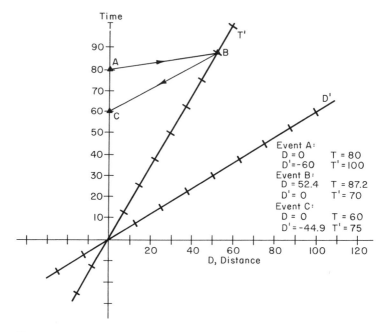

Event A:
D = 0   T = 80
D'=-60   T'=100
Event B:
D = 52.4   T = 87.2
D'= 0   T'= 70
Event C:
D = 0   T = 60
D'=-44.9   T'= 75

FIGURE C.10

cannot exist, the argument goes, because they would undermine our under-standing of causality, a concept so fundamental to our way of thinking that its experimental refutation would render our "logic" and our "common sense" invalid. Meanwhile, the search for tachyons continues—before our models of reality are ready to comprehend them.

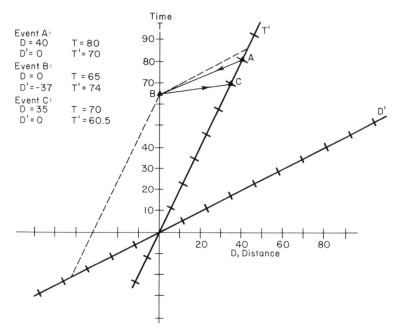

FIGURE C.11

Appendix D

Maxwell's Equations, Electromagnetic Waves, and Special Relativity

D.1 Introduction

At the end of chapter 2 we mentioned that the equations formulated by James Clerk Maxwell served to unify the available experimental facts concerning electric and magnetic phenomena. We went on to point out that Maxwell's equations predict both the existence of electromagnetic waves and their speed of propagation in empty space. The equations further suggest that no matter how the source of the waves might move with respect to an observer in an inertial reference frame, the speed of wave propagation can have only one value, a result that was to become the second postulate of Einstein's special theory of relativity (discussed in section 3.6). We also pointed out in chapter 3, but did not explain, the fact that Einstein's 1905 paper on special relativity bears the title "On the Electrodynamics of Moving Bodies," where "electrodynamics" is that branch of physics dealing with the way that objects are influenced by electric and magnetic forces.[1] It should be clear from all of this that electricity, magnetism, and special relativity are closely related. Yet to this point we have discussed this relationship only briefly and in connection with Einstein's second postulate. It is the purpose of this appendix to clarify the relationship and to explain some of the statements made about Maxwell's equations in chapters 2 and 3.

We have chosen to place this discussion in an appendix because we believe that our primary exposition of special relativity should be kept as straightforward and unencumbered as possible. In particular, we centered our presentation in chapter 3 on Einstein's fundamental considerations of

[1] The term "electrodynamics" was first used in 1825 by André Marie Ampère, for whom the unit of electrical current (the "amp") is named.

space and time—for that is the essence of the special theory. On the other hand, as we will explain presently, Einstein's clue that a reexamination of these cornerstones of physics might be valuable came from electrodynamics (hence the title of his paper). Our discussion here focuses on the *ideas* contained in Maxwell's equations, and so we have sometimes made statements that may be many mathematical steps removed from the equations themselves or that may be particularizations as opposed to general conclusions to be drawn from the equations.

D.2 THE MEANING OF MAXWELL'S EQUATIONS

Maxwell's four equations shown in box D.1 describe the results of centuries of experimentation with electric and magnetic phenomena. We will

BOX D.1

MAXWELL'S EQUATIONS

1. $\epsilon \vec{\nabla} \cdot \vec{E} = d$
2. $\vec{\nabla} \cdot \vec{B} = 0$
3. $\vec{\nabla} \times \vec{E} = -\dfrac{\partial \vec{B}}{\partial t}$
4. $\vec{\nabla} \times \vec{B} = \mu \vec{j} + \mu\epsilon \dfrac{\partial \vec{E}}{\partial t}$,

where $\vec{B}$ describes the magnetic effects,

$\vec{E}$ describes the electric effects,

d measures the electric charge present,

$\vec{j}$ measures the electric current present,

ϵ is a constant of nature, easily measured in the laboratory, which expresses certain electrical properties of space (see the discussion in appendix D.2 for more details),

μ is another constant of nature, easily measured in the laboratory, which expresses certain magnetic properties of space (see the discussion in appendix D.2 for more details).

discuss the meaning of each of these equations in this section. Our discussion could center on the mathematical expressions themselves, that is, on the meaning of the symbols that are used in box D.1, but we will not constrain ourselves to that narrow course.[2] Rather, we choose to focus on the idea behind each equation and we will relate that idea to simple observations that are made of electromagnetic phenomena. In fact, we could carry out the discussion that follows without any explicit reference to Maxwell's equations—and readers who shudder at the sight of mathematical symbols may indeed read this appendix in exactly that way—but the equations provide convenient and explicit references for our discussion. They represent observed facts of nature, and we have to call these facts *something*. We choose to call them by the equations that Maxwell wrote down.

Electric Forces

Electric charges in nature are distinguished by the adjectives "positive" and "negative." Which sort of charge is designated by which adjective is a matter of pure convention. On a dry day if you rub an inflated balloon against the hair on your head, you will create a negative electric charge on the balloon and a positive electric charge on your hair. One could call the charge on your hair negative and that on the balloon positive, but the international convention among physical scientists is to do it the other way around (see figure D.1).[3] It is a well-known fact that if two objects have similar electric charges (both positive or both negative) these objects will repel one another. In contrast, if two objects have opposite electric charges (one positive and the other negative) they will attract one another. One can be more quantitative in analyzing these facts. Careful measurements permitted physicists to determine a law of nature expressing the quantitative relationship between the size of the electric charges on the objects, their distance apart, and the size of the force that they exert upon one another.

---

[2] Although explanations of each of the symbols in the equations can, in fact, be found in the text of this appendix, especially in the footnotes.

[3] The reason for the appearance of the positive and negative electric charges in this situation is not pertinent to our discussion, but it may be of interest to some readers. Briefly, the particles of matter that comprise the material of the balloon and the hair carry electric charges. When the constituent particles of these dissimilar materials are brought into close proximity (by rubbing them together) some of the charges (they are called electrons) tend to leave the hair and cling to the balloon, giving the balloon an overall negative electric charge. The hair, on the other hand, is left with a deficit of negative electric charge (that which now resides on the balloon), and so it is positively charged.

Head of hair

Balloon

FIGURE D.1

This law is called Coulomb's law, and its essence is expressed by the first of Maxwell's equations.[4] But Maxwell did not state the law as a relation between charges and forces as we have just done; instead he used a far more powerful approach (illustrated in figure D.2).

There are two ways of thinking about the forces (of attraction or of repulsion) that charged objects exert on one another as suggested by figure D.2. Using what is called "action-at-a-distance," one imagines that a charged object somehow reaches out across space and exerts a force on another object. This is the usual way school children are taught to think about the force of gravity. But another approach works better for describing electric (and magnetic, gravitational, or any other) forces: one imagines that a charged body somehow *modifies the properties of the space around it*. This

[4] This law is named for the French physicist Charles Augustin de Coulomb (1736–1806). One of the common units expressing the amount of electric charge is called the coulomb. It turns out that the law of electric force is identical mathematically to Newton's law of gravity: the size of the force depends on the product of the two charges involved divided by the square of their separation distance; the size of the gravitational force (as illustrated in box 5.1) depends on the product of the two masses involved divided by the square of the distance separating them.

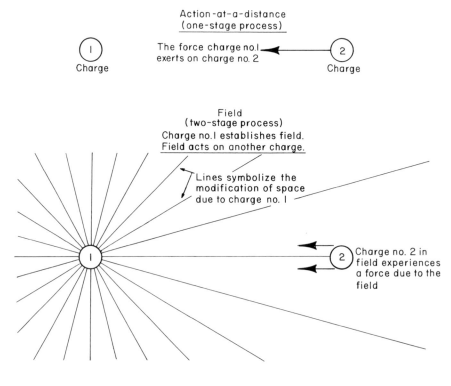

FIGURE D.2

modified region of space is distinguished by saying that there is an "electric field" present in it. A phrase such as "modifies the properties of space" is accurate but requires a careful explanation. We are not talking about anything at all arcane;[5] the idea is as follows. If an electric field exists in a region of space, a charged object will experience an electric force while in that region. As in figure D.2, we picture one charged object establishing a field (modifying space) in its surroundings; a second charge in this same region of space experiences an electric force by virtue of the field's existence. The lines emanating from the first charge in figure D.2 are called

[5] In fact, this way of looking at forces is exactly the one discussed at the end of section 5.7. In general relativity, Einstein states that masses alter the (geometrical and temporal) properties of space, or, more briefly, they alter the properties of spacetime. These alterations give rise to effects that a Newtonian would interpret as a force of gravity.

"field lines," and they are meant to symbolize the fact that space around the charge has been modified.[6]

We say that this is an improvement over the action-at-a-distance approach to describing electric force. In fact, Einstein said of the advent of the field concept:

> This change in the conception of reality is the most profound and fruitful one that has come to physics since Newton.[7]

And yet it seems that we have only made the situation more complex by introducing the intermediate agent of the electric field. It is true that we have added another link to the chain of reasoning, but it turns out that the mathematical treatment of electric (and magnetic and gravitational) phenomena is made more simple and powerful by thinking in terms of fields. A pictorial representation of an electric field surrounding a charged object is shown in figure D.3. In Maxwell's equation 1, $\epsilon \vec{\nabla} \cdot \vec{E} = d$, we find an expression for the geometric nature of the electric field (given by the left side of the equation; the letter $\vec{E}$ is the symbol representing the electric field) established by a charged object (symbolized by the letter d on the right side of the equation).[8] This geometric statement about the electric field turns out to be mathematically equivalent to Coulomb's law of electric force garnered from experiments in the laboratory. We will see presently why the

---

[6] The field lines in figure D.2 can be used to provide a rather detailed "picture" of the electric field. They are also called "lines of force" and physicists often think of fields in terms of these lines.

[7] Einstein, *Essays in Science*, p. 44.

[8] As explained in box D.1, the symbol ϵ (and later on we will see the symbol μ as well) is a so-called "constant of nature"—a number whose value is determined from laboratory measurements. See also the discussion of these numbers in section D.4 below. Small arrows are drawn above some of the symbols in this and subsequent Maxwell equations. An arrow indicates that the symbol beneath represents a quantity whose direction or orientation in space must be considered in the various mathematical operations specified by the equations—these are so-called "vector" quantities. The combination of symbols $\vec{\nabla} \cdot$ on the left side of the equation in combination with the electric field $\vec{E}$ produces a mathematical expression describing the degree to which the lines of force representing the electric field tend to diverge from one point in space. Mathematicians call the symbols $\vec{\nabla} \cdot$ a "divergence operator" for this reason, although Maxwell in his classic (but highly mathematical) work on electricity and magnetism published in 1873 calls this mathematical operator the "convergence"; see James Clerk Maxwell, *A Treatise on Electricity and Magnetism* (New York: Dover Publications, Inc., 1954), p. 30. So Maxwell's equation 1 describes the tendency of the electric field lines to diverge from an amount of charge d located at any point in space. This divergent character of the electric field is suggested by the lines of force sketched in the lower half of figure D.2 and again in figure D.3.

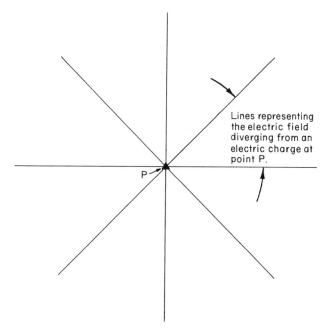

Lines representing
the electric field
diverging from an
electric charge at
point P.

FIGURE D.3

field approach used in equation 1 is so much more powerful than an analysis based on forces acting-at-a-distance.

Magnetic Forces

Let us now discuss some basic magnetic phenomena, which tend to be a bit more difficult than electric ones. In particular many of the figures associated with this discussion must represent on book pages what are actually three-dimensional situations, taxing both our ingenuity at explanation and the patience of readers not used to visualizing things in three dimensions. We begin with the observation that two bars of iron will ordinarily exert no noticeable electromagnetic forces on each other (figure D.4), but they can be treated in various ways to render them "magnetized," meaning that they are capable of exerting and responding to magnetic forces.[9] When this magnetization occurs, the bars become "polarized," meaning that they acquire

[9] There are various ways of magnetizing an iron bar. Probably the simplest is to rub the iron across a magnet. The stronger the magnet, the more quickly the iron bar will become magnetized. Our example employs bars made of iron, but other materials can be magnetized

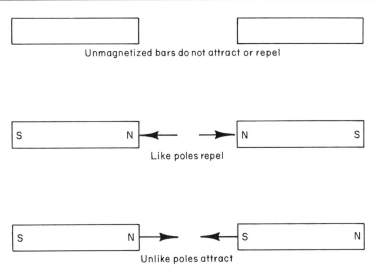

FIGURE D.4

two magnetic poles; the poles are designated "north" and "south" because if either of the bars is suspended at its center by a fine thread, the
bar will swing around until its "north pole" points roughly north in the
geographical sense of the term. Indeed, this is the principle behind the
magnetic compass used (at least since the eleventh or twelfth centuries)
by navigators.[10]

In figure D.4 we have sketched the results of two experiments with bar
magnets. When two similar poles (both north or both south) are brought
close together, they repel each other; when a north and a south pole are
brought close, they attract each other. The compass phenomenon suggested
that the earth itself is a huge magnet with one pole in the north and the other
in the south; the needle of the compass responds to the terrestrial magnetic
forces just as it would in proximity to an iron-bar magnet.

As with electric forces, physicists prefer to deal with magnetic forces in
terms of fields. The region of space surrounding a bar magnet is thought of

---

too (among them nickel and certain ceramics). On the other hand, some materials such as
window glass or water will not become magnetized.

[10] The compass needle is a small magnet able to pivot about its center; its "north pole"
will tend to point in a northerly direction. Notice that we are being careful not to say that the
compass needle or any other bar magnet will swing around to point exactly north. It turns out
that compasses will not, in general, point exactly north, a discrepancy called the "magnetic
deviation," tables of which for various locations on the earth are available to navigators.

as being modified by the presence of a magnetic field that is capable of exerting forces on magnets (figure D.5). Maxwell's equation 2, $\overrightarrow{\nabla} \cdot \overrightarrow{B} = 0$, expresses the geometric properties of these magnetic fields (the magnetic field is symbolized by the letter $\overrightarrow{B}$) just as equation 1 does for electric fields.[11]

Let us summarize our discussion to this point. Maxwell's first equation describes the geometric properties of *electric* fields due to electric charges; it expresses the results of experiments with electric charges by describing how the space surrounding a charge is modified so as to exert forces on other charges—that is, it describes the electric field caused by an electric

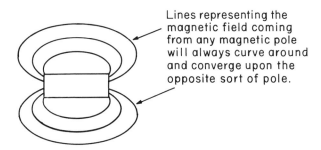

Lines representing the magnetic field coming from any magnetic pole will always curve around and converge upon the opposite sort of pole.

FIGURE D.5

[11] Notice that the left side of equation 2 is identical to the left side of equation 1 except that the letter B now appears where the letter E appeared before (and the constant of nature ϵ does not appear in equation 2). Again, the combination of symbols $\overrightarrow{\nabla} \cdot$ in conjunction with the magnetic field $\overrightarrow{B}$ represents a mathematical expression for the degree to which the magnetic field lines of force tend to diverge from a point in space. But the right-hand sides of equations 1 and 2 are very different. In equation 1 the symbol d stands for the amount of electric charge (positive or negative) giving rise to the field lines, whose tendency to diverge geometrically from the point at which the charge is located is described by the left side of the equation. But there are not known individual "magnetic charges" the way there are individual positive and negative electric charges. In other words, there are no isolated, independent north or south poles of magnetic force; magnetic poles always occur in north-south pairs, and the magnetic field lines leaving one pole will always converge onto the corresponding opposite pole. Isolated poles, or "monopoles," as they are called, have been predicted in various physical theories, but no detection of an actual magnetic monopole has ever been confirmed (unconfirmed reports have appeared from time to time). And so, as illustrated in figure D.5, lines of magnetic force emanating from a magnetic pole will always turn around and converge upon the opposite sort of pole. Compare this illustration of field lines with that of electric field lines in figure D.3. Equation 2 expresses this geometric property of magnetic fields by equating the tendency of the magnetic field lines to diverge away from one another (the left side of the equation) to zero.

charge at any point in space. Similarly, the second equation describes the geometric character of a *magnetic* field.

The Relation between Magnetic and Electric Fields

Maxwell's other two equations express experimentally determined relationships between magnetic fields and electric fields. Equation 3, $\vec{\nabla} \times \vec{E} = -(\partial B/\partial t)$, asserts that if a magnetic field should vary in its strength or direction as time goes on (such variations are described by the right side of the equation),[12] an electric field will be generated in space; the geometric properties of the electric field so produced are described by the left side of the equation.[13] Notice that the notion of a field here takes on a reality *independent* of any electric or magnetic *force*. A changing magnetic *field* (magnetic modification of space) will produce an electric *field* (another sort of modification of space). All of this could be stated in terms of *forces* acting on magnets and charged objects, but such a description would become extremely complex compared to one based on the field concept. Later we will see that fields can be used to predict phenomena (electromagnetic waves) that bear no direct relation whatever to forces acting on objects or to electric charges or magnetized materials.

Figure D.6 illustrates the sort of experiment that prompted formulation of Maxwell's equation 3. The situation pictured is three-dimensional. A meter for measuring the movement of electric charge is connected to a loop of wire. No charges are moving through this wire (there is no battery or other power source connected to the wire), and the meter reads zero. Now

---

[12] This side of the equation is a quotient. The symbol ∂ means "a change in value" of the quantity following it. In mathematics it is called a "partial differential operator." Thus the quotient may be verbalized as follows: "The change in magnetic field divided by a corresponding change in time," or, "the time rate of change of the magnetic field."

[13] The combination of symbols $\vec{\nabla} \times$ in conjunction with the electric field $\vec{E}$ yield a mathematical expression for the way that the electric field varies from point to point in space. In particular, this variation can arise because the field lines bend in space, in which case $\vec{\nabla} \times \vec{E}$ measures the degree of bending. The modern mathematical term for the symbols $\vec{\nabla} \times$ is, appropriately enough, the "curl operator," although Maxwell called it the "rotation" (*Treatise*, p. 30). The field lines produced in electromagnetic waves do not always bend, and in fact they are most easily pictured as straight lines; however, the "curl operator" is still the appropriate device to use in describing how the field changes in strength at various points in space. The divergence operator mentioned in an earlier footnote in connection with Maxwell's equation 1 also measures changes in field strength at different points in space, but those changes refer only to changes *along the direction of the field lines* themselves. The curl measures changes in field strength in directions *not* along the field lines.

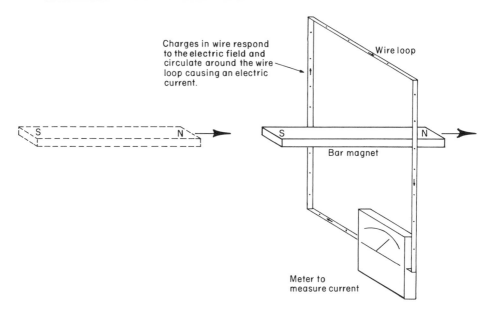

Charges in wire respond
to the electric field and
circulate around the wire
loop causing an electric
current.

Wire loop

S   N

S   N

Bar magnet

Meter to
measure current

FIGURE D.6

bring a bar magnet close to the wire loop and move the magnet around rapidly with respect to the wire. In fact, the experiment gives the most dramatic results if the magnet is passed back and forth through the loop. The needle of the meter will move, indicating the presence of moving electric charges and hence an electric field in the wire. Why? The magnet establishes a magnetic field in the space surrounding it. As the magnet is moved around, its associated magnetic field moves too: the field changes its direction and its value at each point in space. In particular, the magnetic field changes at each point in space occupied by the wire loop. Because the magnetic field is changing in time, Maxwell's equation 3 predicts that an electric field will be established in space, and the electric field so generated exerts forces on charged particles in the wire and causes them to move around (producing an electric current in the wire). This is the principle used in electric generators in automobiles, bicycles, or commercial power plants.

So the content of Maxwell's equation 3 may be summarized by saying that a magnetic field changing in time produces an electric field in the same region of space. This electric field will disappear as soon as the *change* in the magnetic field ceases.

Maxwell's equation 4, $\vec{\nabla} \times \vec{B} = \mu\vec{j} + \mu\epsilon\,(\partial\vec{E}/\partial t)$, expresses the fact that an electric field that changes its strength or direction in time will produce a magnetic field in the surrounding space. This is a symmetric counterpart of equation 3. Again, the left side of the equation expresses the geometric properties of the magnetic field produced, and the right side expresses the variation of the electric field in time.[14] But there is something else in equation 4. If a flow of (either positive or negative) electric charge (an electric current) exists, then a magnetic field will also be established as a consequence of the charge motion. A simple experiment involving four compass needles and a length of straight wire is shown in figure D.7 to illustrate this fact. Again, the figure represents phenomena in three dimensions. The compass needles are used to detect magnetic fields; they are arranged to lie in the book page at equal distances from the wire, as shown in the figure. The wire passes perpendicularly through the page so that all we see of it in figure D.7 is its circular cross-section. In the first part of the experiment (the top portion of figure D.7) the wire carries no electric current, and the compass needles assume their usual north-south orientation because of the earth's magnetic field. Now suppose that a large current flows in the wire. The compass needles will swing around so that they line up along the circumference of a circle concentric with the wire; they no longer line up in the earth's magnetic field.[15] In other words, as long as charges move in the wire a magnetic field is established in the vicinity of the wire. Again, equation 4 specifies the configuration of this magnetic field in space; the right side contains a letter (j), which represents the rate at which charges are flowing in that region of space.

So according to equation 4 there are two things that can establish a magnetic field in space (corresponding to the two terms in the right side of the equation): an electric field that varies in time, or a flow of electric charge (an electric current).[16]

---

[14] Footnotes 12 and 13 describe the various mathematical symbols used in this equation.

[15] The earth's magnetic field still is present, of course, but we use a "large current" in the wire so that the magnetic field produced by this current completely overwhelms the weak terrestrial field. Readers may wish to try this experiment using a magnetic compass to detect the magnetic field from a wire carrying a current. The sort of current moving through the cords of electrical appliances will not do; that is an "alternating current" that changes its strength and direction sixty times each second. A "direct" current is needed for this magnetic experiment, and one may be established by connecting the two metal ends of a length of wire to the two terminals of a flashlight dry-cell battery.

[16] The magnetic field produced by a bar magnet can be attributed to the circulation of electric charges (in other words, to minute electric currents) within the molecular structure of the magnet material. In a sample of unmagnetized material, these currents move with more or less random directions so that their associated magnetic fields cancel one another. The proc-

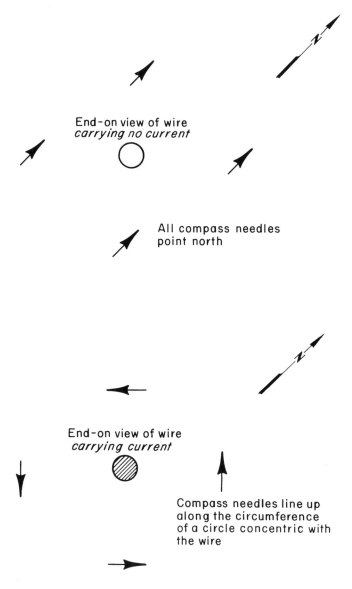

End-on view of wire
carrying no current

All compass needles
point north

End-on view of wire
carrying current

Compass needles line up
along the circumference
of a circle concentric with
the wire

FIGURE D.7

Magnetic Forces Acting on Moving Charges

There remains one more experimental fact of significance in our discussion of Maxwell's equations, and Einstein's 1905 paper showed that the equations contained the fact implicitly. The fact is this: if a charge moves with respect to a magnetic field, the charge *may* (depending upon its direction of motion) experience a force due to that magnetic field. A charge at rest in a magnetic field will experience no force due to that field. This phenomenon is illustrated in figure D.8, which must be interpreted as another three-dimensional image.

The directions in three-dimensional space represented by figure D.8 are very important to the following discussion, and so we want to define the terminology we will use in talking about them. The two directions in space that lie *in* the book page as you see it we will call the ''right-left'' and the ''up-down'' directions. The third direction *perpendicular* to the book page we will call the ''in-out'' direction. Suppose we use a long bar magnet lying in the ''right-left'' direction to modify a region of space with a magnetic field. We know that another magnet in this region of space will experience a force due to this magnetic field, but experiments show that *moving electric charges* can also experience a force due to this *magnetic field*. Whether or not a moving charge experiences a force due to the magnetic field depends on the direction of the charge's motion with respect to the field. Suppose a charged object is moving through space in the ''right-left'' direction (that is, parallel to the bar magnet's length) as shown in figure D.8c. We would continue to observe the charged object moving uniformly; in other words, no forces act on the charge (we are assuming here that there are no electric fields around to create an electric force on the object and that the force due to gravity is negligibly small).[17] But now suppose the charged object in figure D.8 has been made to move in the ''in-out'' direction (that is, perpendicular to the page) as shown in figure D.8a. In that case the object will experience a force in the ''up-down'' direction (that is, toward or away from the bar magnet; whether the force is toward or away from the magnet will depend on whether the charge is positive or negative). And if the charge moves in the ''up-down'' direction (toward the bar magnet as in

---

ess of magnetizing a material according to this view is one of aligning the minute electric currents so that their individual magnetic fields reinforce one another.

[17] We must also assume that the charge moves near the middle of the bar magnet, far from its ends. Near the ends of the magnet the magnetic field curves around to enter the magnetic pole, and in that case the charge moving left-right will experience a magnetic force.

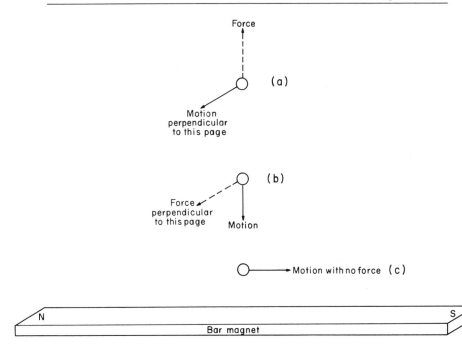

figure D.8b), then it will feel a force in the "in-out" direction (perpendic-ular to the page of the book; again, whether the force acts into or out of the page depends on whether the charge is positive or negative).

So if the charged object does not move at all, it experiences no force due to the magnetic field; if it moves in a certain direction ("right-left" or along the length of the bar), it also experiences no magnetic force; if it moves in any direction but "right-left," a force acts on the charge due to the mag-netic field, and the force is perpendicular to the direction of the charge's motion.

This complex situation is difficult to illustrate in a book and to discuss without benefit of a full three-dimensional model. The forces experienced by the charge depend on its state of motion: the faster it moves with respect to the magnet, the greater the force; the force arises only when the object moves in certain directions with respect to the magnet; and the direction of the force is perpendicular to the direction of the charge's motion. These are the observed experimental facts, and while our illustration has involved the magnetic field established in space by a bar magnet, it turns out that similar sorts of arguments apply to charges moving in any sort of magnetic field.[18]

To summarize, in general a charged object can experience a force due to two sorts of fields (again we are neglecting gravity): it will experience a force due to an *electric* field, regardless of the charge's state of motion, and it will experience a force due to a *magnetic* field that depends critically on the charge's state of motion. In particular, if the charge is not moving with respect to the magnetic field, there will be no magnetic force.

D.3 ELECTROMAGNETIC WAVES

We will now concentrate our attention on Maxwell's equations 3 and 4 because they describe the interrelatedness of magnetic and electric phenomena. We first succinctly rephrase the ideas inherent in these two equations for a special situation: a volume of totally empty space. With no material objects immediately present, there can be no bodies carrying electric charge (and hence no electric current). But there can still be electric and magnetic fields (due to remote electrically charged or magnetized objects, or, as we will see presently, due to electric and magnetic fields created by atomic processes in distant matter). Here is our paraphrasing of equations 3 and 4 for a volume of empty space:

> *A magnetic field that varies in time produces an electric field that varies in time* (equation 3).

> *An electric field that varies in time produces a magnetic field that varies in time* (equation 4).

The symmetry inherent in these two statements is striking. Interchanging the words ''magnetic'' and ''electric'' in the first statement produces the second one. This statement of Maxwell's equations for empty space clearly suggests the possibility of creating a natural ''electromagnetic'' perpetual motion:

> An electric field that varies in time produces a magnetic field that
>   varies in time,

---

[18] The force experienced by a moving charged object in a magnetic field is sometimes called the ''Lorentz force,'' although it was first expressed mathematically in 1889 by O. Heaviside. See Whittaker, *A History of the Theories of Aether and Electricity*, I, 310.

which produces an electric field that varies in time that produces a
    magnetic field that varies in time,
which produces an electric field that varies in time . . .

This cyclic self-generation of time-varying electric and magnetic fields is predicted in quantitative detail when Maxwell's equations are combined (using the techniques of calculus); the solution of the resulting combined equation represents an undulatory motion of the magnetic and electric fields through space, and it is this undulation that constitutes what is called an "electromagnetic wave."[19]

This is the background to the prediction of electromagnetic waves from Maxwell's equations. Without the field concept this prediction would have been much more difficult; notice that no forces—in fact, no matter upon which forces can act—enter the discussion at all. The waves result from the natural properties of two different sorts of *fields*. But how do we visualize these waves? Visualization is possible using the field concept (another illustration of the great advantage that accrues to the user of fields as opposed to one who insists on dealing only with the electric and magnetic forces and action-at-a-distance), and we present an appropriate picture in figure D.9. We make no attempt to explain all of the details of this drawing. The details emerge from a comprehensive mathematical analysis of Maxwell's equations, and that level of treatment is not appropriate to this book. The figure shows a "snapshot" of an electromagnetic wave propagating to the left along the line labeled "direction of travel" in the figure (that is, we have "frozen time" to look at the pattern of electric and magnetic fields in space at one instant; in reality, the whole pattern moves along the direction of travel to the left as time goes on). There are really two wave components to this figure, corresponding to the two root words in "electro-magnetic." The electric field at several points in space along the direction of travel is shown by the solid vertical arrows. These arrows represent the electric field in that each represents the direction of the electric force that would be experienced by a positive electric charge if placed at the base of the arrow. The length of each arrow represents the magnitude of that electric force. The sinuous curve passing through the tips of the arrows helps one to visualize what the lengths of the arrows would be at all of the other points in

---

[19] Whittaker, *A History of the Theories of Aether and Electricity*, I, 194, indicates that Michael Faraday, in a publication of 1851, hinted at the possible electromagnetic nature of light some time before Maxwell's work.

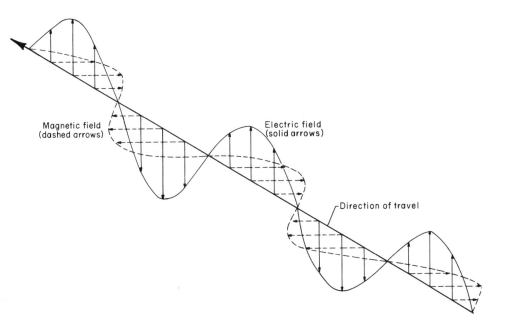

Magnetic field (dashed arrows)

Electric field (solid arrows)

Direction of travel

FIGURE D.9

space along the direction of travel where we have not—for reasons of clarity in the figure—drawn electric field representations. Similarly, the dashed arrows are meant to represent on a page what is in reality a three-dimensional phenomenon: the dashed arrows should stick out at right angles to the solid arrows representing the electric field. The dashed arrows represent the magnetic field at various points in space along the direction of travel. The direction of the lines is the direction along which a compass needle would line up if placed at the point at the base of each arrow. The length of the lines represents the magnitude of the force that the magnetic field would exert on a magnet or a moving electric charge at that same point.

We repeat, this picture represents the spatial arrangement of the electric and magnetic fields in an electromagnetic wave. In a "movie," as opposed to our snapshot, this entire pattern would move to the left at the speed of light so that each point in space along the direction of travel would experience undulatory electric and magnetic fields as time goes on, each field increasing, reaching a maximum and then decreasing to a minimum, over and over again in a regular cycle.

Why Waves?

We have tried to emphasize that Maxwell's equations summarize the results of experiments designed to investigate the properties of electric and magnetic phenomena. We then stated that a mathematical combination of Maxwell's equations results in the prediction of electromagnetic waves of the sort sketched in figure D.9. Without burdening our discussion with any mathematics let us formulate a qualitative answer to the question why these waves result from combining Maxwell's equations. A changing magnetic field can produce a changing electric field and vice versa. To attain a truly self-sustaining (perpetual) electromagnetic phenomenon, we must therefore seek some regular (as opposed to erratic) time variation in both electric and magnetic fields. Regularity in the variations is suggested by the symmetry of the equations: a regular variation in electric field should produce a similarly regular variation in the magnetic field, and the combination of these regular variations should generate a self-sustaining electromagnetic phenomenon. Any sort of regular and perpetual time variation of the field strengths other than one increasing and then decreasing in strength (that is, an undulatory variation) would imply a field that changes in a regular way by growing larger and larger; to be perpetually changing such a field would have to become infinitely large, something that would require, somewhere, an infinite electric charge or magnet. This requires, in turn, some infinite source of energy to create these infinite fields. We seem compelled then, in order to avoid the requirement of an infinite energy resource, to think in terms of fields that vary in time by increasing and then decreasing in strength according to some regular pattern—an undulatory variation. The undulations shown in figure D.9 are the simplest such undulations from a mathematical point of view. We could have drawn more complex ones (and such electromagnetic waves can be created in nature), but these simple ones are easiest to understand and they do represent, in fact, the usual sort of light waves that are created naturally.

The Creation of Electromagnetic Waves

We have described how Maxwell's equations predict the self-perpetuation of electromagnetic waves, but we have not yet said how such waves are created in the first place. In one common method, electric charges in a wire are pushed back and forth rapidly so that the electric field associated with the charges changes rapidly in time. This time-varying electric field produces a time-varying magnetic field, which starts the electromagnetic wave propagating through space. The charges can be moved back and forth in a device called an ''antenna'' found at commercial broadcasting facilities. But an antenna is not necessary for the production of electromagnetic waves.

The atoms comprising all matter are also capable of creating electromagnetic waves. When atoms are given some excess energy (by heating them for example), they can spontaneously give off some of that energy by emitting short bursts of electromagnetic waves. This is the origin of the electromagnetic waves in electric lighting, in the sun, in fireflies, and in the flashbulbs used by Pablo, Gertrude, and Alice along our railroad track.

D.4 THE SPEED OF LIGHT

As we remarked in section 2.7, Maxwell's equations do more than predict the existence of electromagnetic waves. They also predict the speed at which the waves move through space, a speed that enters relativity theory in such a fundamental way that we wish to address two related questions concerning it: the actual value of the speed predicted by Maxwell's equations, and the fact (postulated by Einstein in his paper of 1905) that the measured speed of light is independent of the relative motion of the light source and of inertial observers. Let us address these questions in order.

The Predicted Value of the Speed of Light

Look again at Maxwell's equations in box D.1. There are two so-called ''constants of nature'' in these equations that measure the electric and magnetic properties of the medium in which the fields exist.[20] Both of these numbers are measured easily in a physics laboratory, and their values for various materials are available in physics handbooks.

When Maxwell's equations are combined as described previously to pre-

---

[20] These two constants of nature measure the attenuation in strength of magnetic (μ) and electric (ϵ) fields due to their passage through a medium. For example, suppose two electrical charges are separated by a fixed distance, and some device is attached to them to measure the force with which they attract one another. The device registers some value for the force when the charges are in a vacuum; but it will register a smaller force when the same two charges are located in oxygen and a still different and much smaller value when they are in a chamber filled with water. In other words, the medium surrounding the charges has attenuated the strength of the electrical force compared to its value in a vacuum. This attenuation is meas-

dict the existence of electromagnetic waves, the result also predicts the speed for those waves. That speed is found by taking the square root of the product of ϵ and μ,

$$C = \sqrt{\epsilon\mu}.$$

When values of ϵ and μ for empty space (nearly the same values for the numbers are obtained for air) are substituted in this square root, the speed of the waves comes out to be just that measured for light.

Because ϵ and μ depend on the electric and magnetic properties of the material through which the fields pass, the speed of light will also differ for different substances. But the same general formula, combined with laboratory measurements of ϵ and μ, can be used to predict that speed; the predictions are substantiated by laboratory determinations of the speed of light in various substances.

Einstein's Second Postulate

Next, we discuss Einstein's second postulate that the measured speed of light in space will be the same for all inertial observers. To understand the profound contrast between electromagnetic waves and other types of waves, let us first consider a mechanical wave on water (we could equally well choose other mechanical waves such as sound waves or waves on a string). Such waves travel through the water medium at a well-defined speed, a speed that depends on the properties of the water. No matter how fast the source of those waves travels through the water (for example, we can imagine that the waves are caused by the prow of a boat moving through the water), the speed of the water waves is independent of the speed of the wave source; the wave speed is determined only by the mechanical properties of the medium. So far this would seem consistent with Einstein's second postulate. Light waves, too, travel with a speed that is independent of the speed of the source through space, as though there were some medium that determines that speed independently of the motion of the source. Indeed, as we remarked at the end of chapter 2, for years physicists sought to determine the properties of this putative medium, called the "luminiferous ether," but to no avail. Maxwell's equations do not at all rule out the existence of such a medium, and so the search for ether was not un-

---

ured by ϵ. Similarly, two magnetic poles will exert a mutual magnetic force whose strength depends on the medium surrounding the magnets. The number μ measures the attenuation of the force from the value it would have if the magnets were in a vacuum.

reasonable. But Maxwell's equations can also be interpreted in a very different way that allows for (what is now recognized to be) a marked contrast between electromagnetic waves and waves of other sorts.

Imagine some water waves moving across the surface of an otherwise calm lake. An observer in a motorboat is moving through the water in the same direction as the waves. If the motorboat moves with respect to the water at a speed equal to the speed of propagation of the water waves, then the water waves will appear to stand still relative to the observer in the motorboat. Looking out at the lake from the speeding boat, the observer sees stationary "indentations" in the surface of the water. By moving at an even greater speed with respect to the water, the observer in the motorboat can even outrun the waves, eventually leaving them completely behind. But as we have remarked in chapter 3, one cannot overtake light waves emitted in space. Light waves will always appear to approach an inertial observer at the same speed. The speed of electromagnetic waves is determined by the electric and magnetic properties of empty space (properties quantified by the constants of nature ϵ and μ); its speed is defined with respect to those properties and not with respect to any medium. The speed of the observer or of any medium is not called for in Maxwell's equations because such speeds are irrelevant.

Let us state this again, but with some clarifying detail. In the case of mechanical waves such as water waves or sound waves, the speed of propagation is calculated from an analysis of the mechanical properties of the medium through which the wave travels (for example, one uses the mass or inertia of some small piece of the material medium and the strength of the forces that act between contiguous particles of the material). Therefore, "the speed" of the wave is a perfectly meaningful term; speed must always be referred to some frame of reference, and here the frame of reference must be the medium with respect to which the wave motion is taking place. Of course an observer is able to move herself with respect to that medium (through the air carrying the sound wave or through the water carrying the water wave), and the speed that the observer measures for the wave will depend on her motion with respect to the medium as well as on the calculated propagation speed of the wave with respect to the medium. If the observer moves in the same direction as the wave and at a speed equal to that of the wave, the wave will stand still with respect to that observer (the wave ripples will not move across the water but will be stationary indentations in the water surface as discussed above).

But unlike mechanical waves such as sound or water waves whose speed is determined by the inertia, internal forces, and other characteristics of the mechanical wave medium, electromagnetic waves consist of electric and magnetic fields that exist in *empty space with no medium present at all.* And since electromagnetic waves are self-propagating by virtue of the properties of electric and magnetic fields (as expressed by Maxwell's equations), no medium need be present for electromagnetic waves. The propagation speed of these waves is determined by two constant numbers that can be measured for empty space in the laboratory. These alone suffice to calculate the speed of light. We emphasize the words ''the speed'' because we must pause to ask what this term means. ''The speed'' with respect to what? In the case of mechanical waves ''the speed'' was clearly determined with respect to the mechanical medium carrying the wave: but with light in empty space, what can this term mean? Maxwell's equations suggest that any inertial observer will calculate (and measure) the same speed for light since the numbers involved in its calculation do not refer to any medium or to any state of motion of the observer.

A consideration similar to this was evidently of key importance in Einstein's development of relativity theory. This consideration, according to Einstein's autobiographical notes, took the form of

> a paradox upon which I had already hit at the age of sixteen: If I pursue a beam of light with the velocity c (velocity of light in a vacuum), I should observe such a beam of light as a spatially oscillatory electromagnetic field at rest.

In other words, if one moves along at the same speed as a light wave, the wave should appear to be at rest with respect to the observer, just as the water wave appears as a fixed (and ''spatially oscillatory'') indentation on the surface of the water to someone moving though the water with the same direction and speed as the water wave. Einstein goes on:

> However, there seems to be no such thing [electric and magnetic fields at rest in an electromagnetic wave], whether on the basis of experience or according to Maxwell's equations.[21]

In other words, an observer just cannot move along with a light wave; to do so would require the pattern of electric and magnetic fields to be at rest with respect to this observer; but Maxwell's equations tell us that the essence of

---

[21] Einstein, ''Autobiographical Notes,'' in Schilpp, *Albert Einstein: Philosopher-Scientist,* I, 53. This essay is highly technical in places but is quite suitable for general readers in others.

light is the self-perpetuation of the magnetic and electric fields, and this self-perpetuation, in turn, depends on their *variations in time*. Electric and magnetic fields in a light wave cannot be at rest with respect to any observer. The suggestion is clear, then, that one simply cannot move with a wave of light. Furthermore, no matter how one is moving in space with respect to any inertial frame, Maxwell's equations predict the same speed for light. The equations make no connection with any reference frame.

This way of analyzing Maxwell's equations in itself does not *prove* Einstein's postulate about the constancy of the speed of light for all inertial observers, nor does it answer many other significant questions about the peculiar properties of electromagnetic waves: but it suggests strongly that the postulate is a correct representation of nature.

D.5 The Unity of Electric and Magnetic Forces and the Title of Einstein's 1905 Paper on Relativity

At the end of section D.2 we described the magnetic forces that can act upon charged particles moving with respect to a magnetic field. We emphasized that the charged particles experience a *magnetic* force only when they are in motion with respect to the magnetic field. But magnetic fields, in turn, can be established by moving electrical charges (for example by a current of electric charge as described in Maxwell's equation 4; see section D.2). The relative motion of electric charges and the phenomenon of magnetic forces are therefore closely related.

This relationship intrigued Einstein in the years before 1905, and it appeared to him that something was missing form Maxwell's model "as usually understood at the present [1905] time."[22] If a *charge* were moved near a magnet at rest on a table, one analyzed the situation by calculating the resulting "magnetic force" on that charge (the force illustrated in figure D.8). But if the same charge were held at rest on the table and *the magnet* moved, the analysis suggested by Maxwell's equations differs. When the magnet moves one ignores any consideration of the "magnetic force" and instead invokes Maxwell's equation 3. The moving magnet causes a changing magnetic field, which, by equation 3, creates an electric field near the charge. The charge responds to this electric field by experiencing an electric force. But what difference should it make whether the charge is at rest

[22] Einstein, *The Principle of Relativity*, p. 37.

on the table and the magnet moves or vice versa? Only the relative motion of the charge and the magnet should count. Traditionally the assumption that the charge moves yielded an analysis in terms of a magnetic force; the assumption that the magnet moves yielded an analysis in terms of an electric force. But to Einstein a phenomenon brought on by *any* relative motion on the charge and the magnet had to be described by the same fundamental analysis regardless of whether the charge or the magnet were at rest on the table. That the usual approach to the analysis of moving charges and magnets through Maxwell's equations led to the action of two distinct forces depending on which object sits at rest on the table therefore presented an important puzzle to Einstein.

This explains the title of the 1905 paper on relativity: "On the Electrodynamics of Moving Bodies." As we have explained in section D.1, "electrodynamics" is the study of electric and magnetic forces, and in this paper Einstein was concerned with explaining the way that electric and magnetic fields act on moving charges and especially how these various actions are interrelated. Einstein elucidated that interrelationship in the pages following his development of special relativity and the Lorentz transformations by demonstrating the fundamental identity of electric and magnetic phenomena.

The recognition of this identity resulted from a mathematical analysis of Maxwell's equations and the Lorentz transformations, which we cannot follow here because that would involve us in calculus. The idea behind the analysis is this. The Lorentz transformation equations can be used to calculate how any inertial observer will measure the distances and times for events in spacetime. One is also able to combine these equations to calculate how *forces* experienced by any inertial observer would be experienced by any other inertial observer. Suppose, for example, that we have an electric field in a certain region of space. We know just how a charge at rest with respect to the field will respond to it. We can use the Lorentz transformations to see how this same field will be experienced by a charge that is moving with respect to the field. We find that the moving charge will experience two forces due to the electric field according to the Lorentz transformation analysis: one force, an electric force quite similar to the one it would experience at rest with respect to the field, and a second force *identical in its mathematical form to the magnetic force acting on a moving charge.* In short, Einstein showed that what had been called the "magnetic force" acting on a moving charge can be regarded as an electric force—but an electric force that is now altered by virtue of the relative motion of the electric field and the charged object. Just as measurements of time and distance depend on the relative motion of inertial observers, so measurements of force will depend on the relative motion of inertial observers. An observer riding on an electric charge at rest with respect to an electric field experiences only an electric force in the direction of the field (as discussed in section D.2). A second observer riding on a charge in motion with respect to the same electric field will experience two forces: one very similar to the usual electric force and a second one identical to what had been formerly called a "magnetic" force. In this way, electric and magnetic forces acting on moving charges became unified by special relativity, and the missing (unifying) element in Maxwell's equations was identified. Whether one moves a charge near a magnet at rest on a table or vice versa, the same sort of force acts and the physical analysis is the same.

And here is one final point. As Einstein developed his special theory of relativity, he found that Maxwell's equations are completely consistent with the principle of relativity but that Newtonian mechanics is not. Using what were to be recognized later as the Lorentz transformations, Einstein showed that all inertial observers would agree on the validity of Maxwell's equations for describing electromagnetic phenomena, as required by the principle of relativity. But whereas Maxwell's equations could survive special relativity, Newton's laws could not. Inertial observers could not agree that Newton's laws describe the behavior of objects. What lay before Albert Einstein at that point was the necessity of reconstructing the science of mechanics. Maxwell's equations could remain intact.

Glossary

We have assembled here the technical terms used in this book that may cause confusion. We have written a definition or explanation for each and have supplied explicit references to sections where these terms are introduced and discussed so that the glosses may be supplemented.

Acceleration. This measures the rate at which the velocity of an object changes in time. Velocity is a characterization of motion involving specification of both the speed *and the direction* of motion. Acceleration therefore involves the time rate of change of both speed and direction of motion. See section 2.5 and figure 2.6.

Aristotelian Cosmology (or "Aristotelian system"). A model of the cosmos with the earth at rest at the center; all the celestial objects are in motion about the central earth. This may be the system of cosmology used for the greatest length of time in Western civilization. The term is often used interchangeably with "Ptolemaic cosmology," although Aristotle's model and Ptolemy's were quite different in detail. (An excellent discussion of the various cosmological systems at a fairly detailed level but one still accessible to the general reader is Kuhn, *The Copernican Revolution*.) See section 1.3 and figure 1.1.

Atom. In modern physical sciences this term describes the smallest particle of a given sort of chemical element. A useful model of the atom may be found in an analogy to the heliocentric solar system: the sun is at the center and the planets revolve about the sun. An atom (in this model) consists of a central nucleus around which orbit much less massive electrons. The nucleus, in turn, is made of a combination of "nuclear particles" called protons and neutrons, with the number of protons in the nucleus determining the sort of chemical element represented by the atom.

Axes (in graphs). These are the horizontal and vertical scales drawn on a graph to locate points. They are also called "coordinate axes" and the numbers associated with them are often called "coordinates." The mathematical terms "ordinate" (for the vertical axis) and "abscissa" (for the horizontal axis) are also used.

CAUSALITY. The cause-and-effect relationship; also the doctrine asserting that a cause-and-effect relationship holds.

CLASSICAL PHYSICS. This term is antonymous with ''modern physics'' and suggests physics obtained from Newtonian principles, but not necessarily only that physics known to Newton. For example, Maxwell's equations of electromagnetism are termed ''classical'' physics. The term is used to distinguish between physical science involving relativity theory or quantum theory and physical science that does not use these ''modern'' physical theories.

COLOR (of light). Light is electromagnetic radiation and may be thought of as waves of electricity and magnetism. The color of the light is determined by the ''frequency'' of vibration of these waves (that is, the number of crests striking the eye per second), or, equivalently, by the distance in space between successive crests of the wave (a quantity called the ''wavelength''). In this model, the higher the frequency, the more blue the light. According to another model of electromagnetic radiation, light consists of particles. The color of the light is then determined by the energy of the particles. The more blue the light, the higher the energy. See box 2.1 for a summary of terms used in describing light and other forms of electromagnetic radiation.

CONTRACTION (sometimes ''FitzGerald Contraction''). This term is used to describe the length contraction phenomenon of special relativity (see the discussion in section 3.8). George Francis FitzGerald (1851–1901) was an Irish physicist who, before Einstein's relativity paper was published, proposed that the length of an object might contract as it moves though the ether. He was attempting to explain why experiments had failed to detect the motion of the earth through the ether (especially the Michelson-Morley experiment). H. A. Lorentz elaborated FitzGerald's idea in the years preceding 1905, and so the contraction phenomenon is also sometimes called the ''Lorentz-FitzGerald contraction'' or even the ''Lorentz contraction,'' although FitzGerald was first in proposing the idea.

COORDINATES. In graphs, these are the numbers used to locate the position of a point on the diagram. The number of coordinates needed to specify a location is identical to the dimensionality of the diagram (two numbers in a two-dimensional diagram, three numbers in a three-dimensional diagram, and so forth).

COPERNICAN MODEL of the cosmos. This is the heliocentric model of the
cosmos proposed by Copernicus. The term is also applied as a generic
term for "heliocentric models," although the heliocentric model favored today (due primarily to the work of Kepler) bears slight resem-
blance to Copernicus's model. The term "Copernican revolution"
has been applied to the uproar created by the advocates of the heliocentric model in the seventeenth century. See our discussion in section 1.4.

COVARIANT or covariance. Covariant relations in physical science are
ones that do not depend on the coordinate system used to describe
them (see the discussion in section 5.9). Tensors are mathematical entities that by their nature are independent of coordinates, so that when
a law of physics can be expressed as a tensor relation the law is de
facto covariant. This was used by Einstein in formulating his general
theory of relativity where he wished to express laws that are valid in
any and all coordinate systems.

CURVATURE of space. This is a very tricky (even treacherous) term. It is
often used in books on relativity and often misunderstood or misinter-
preted by readers. We have usually avoided it because we feel that it
is "loaded." It refers to non-Euclidean geometry; that is, wherever
the rules of Euclidean geometry do not hold, one can say that space is
"curved." In particular, geodesics are no longer straight lines in the
Euclidean sense. See section 5.6.

CYCLOTRONS. These devices accelerate charged particles to very high
speeds. The accelerated particles are then permitted to "collide" (or
"interact") with other particles, and from data obtained in the collision one can test models of the structure of matter. The colloquial term
"atom-smasher" used to be applied to devices of this sort. The generic term is "particle accelerator," and other types include van de
Graaf machines, Betatrons, Bevatrons, Synchrotrons, and linear accelerators.

DEFLECTION of starlight. According to the general theory of relativity,
matter changes the geometry of surrounding space from the Euclidean
properties it has in the absence of any matter. Light follows geodesics
(the shortest path between two points in space), which are straight
lines in Euclidean space and curved lines in the non-Euclidean space
near matter. Hence, the path of light rays should bend or be deflected
near matter. This prediction of general relativity theory was first borne

out in observations (in 1919) of starlight passing near the mass of the sun. See section 5.6.

DILATION of time. This is the name given to the slowing of moving clocks predicted in special relativity theory. See section 3.7.

DIMENSION. This term has been confused in some writing about relativity theory. In physics a dimension is a number used to specify some property characterizing an object. For example, two numbers (dimensions) are needed to specify uniquely the location of an object on a surface. In space, one needs at least three numbers (dimensions) to locate an object. If one wishes to ascribe other properties to the object (speed, or temperature, or age) one can add as many dimensions as needed to complete the specification. The "dimensionality" is the number of dimensions needed to specify all of the desired properties of an object. See the discussion in section 3.11.

ELECTRODYNAMICS. This is the branch of physical science dealing with electric and magnetic forces and their effect on the motion of matter. See section D.1.

ELECTROMAGNETIC WAVES. According to Maxwell's equations, a magnetic field that varies in time will produce an electric field that also varies in time. The equations also predict that an electric field that varies in time will produce a magnetic field that varies in time. Thus a perpetual motion of sorts is possible with time-varying electric fields generating time-varying magnetic fields that generate time-varying electric fields and so forth. The equations show that this phenomenon manifests itself as waves that move at the speed of light in space. Indeed, these waves have all of the properties observed for light. Box 2.1 shows the names assigned to various sorts of electromagnetic waves. The term "electromagnetic radiation" is also commonly used. See sections 2.7 and D.3.

ELECTROMAGNETISM, laws of. These are summarized in Maxwell's equations (which describe the interrelatedness of electric and magnetic fields and their origin in electric charges) and in another equation describing the force exerted on charged matter by electric and magnetic fields (the so-called "Lorentz force"). See section D.2.

ELECTRONS. One of the "subatomic" particles or "building blocks" of atoms. These particles can be pictured as orbiting about the central nucleus of the atom. Each bears a negative electric charge.

ELEMENT, chemical and Aristotelian. *Chemical*: All matter can be broken into constituent elements, each comprised of a unique sort of atom. Thus the substance water is described by the well-known formula H_2O; this is chemical symbolism for the fact that water consists of two atoms of the element hydrogen (H_2) and one atom of the element oxygen (O). Ninety-two different elements occur in nature, and several more may be "manufactured" in the nuclear physics laboratory by forcing a collision between atoms of one of the "natural elements" and other particles that have been accelerated to move at very high speeds. *Aristotelian*: According to Aristotelian physics, all matter of the universe consists of five elements. Four of these (fire, air, earth, and water) are found on earth and throughout the sublunary sphere (the volume within the sphere in space bearing the moon). The fifth element, the quintessential element, or ether, exists beyond the sublunary sphere, and the perfect and perpetually moving heavens owe their properties to it. See section 1.3.

ELLIPTICAL. Kepler discovered that the orbits of the planets follow paths in space that are ellipses; Newton subsequently showed that this is a necessary consequence of the validity of his models of motion and of gravity. An ellipse is a symmetric oval; the departure of the oval from pure circularity is measured by a number called the "eccentricity." An ellipse of zero eccentricity is a circle; an eccentricity of one indicates a completely flattened circle, that is, a straight line. See Cohen, *The Birth of a New Physics*, pp. 131–134, for a nontechnical discussion of the properties of ellipses.

ENERGY. In physics this is defined as the capacity to do work, that is, the capacity to exert a force on an object and move it. The energy may be "stored" in a system (by virtue of the arrangement of its parts in spacetime or by virtue of the energy content of the matter present), or it may be energy due to the motion of an object.

EQUABLE. As used in Newton's laws the word means "unvarying" or "uniform."

EQUIVALENCE, principle of. This was Einstein's postulate that in a small volume of space the observed effects brought about by gravity are equivalent to those brought about by accelerating the reference frame in which the observations are being made. See our discussion in section 5.3.

ETHER (sometimes spelled "aether"). In the Middle Ages this was the "quintessential" element comprising the perfect supralunary region of the cosmos. In the nineteenth century the term was applied to the medium thought to carry light waves and to pervade all space. There was never substantial evidence for the existence of the ether except for the fact that physicists could not imagine wave motion (a characteristic that light was known to display) without some medium. Hence its existence was postulated.

EUCLID's rules of geometry; EUCLIDEAN. These are the rules of geometry usually taught in school and they are valid only when the region of space considered is "Euclidean." Einstein's general relativity theory showed that matter can cause space to become "non-Euclidean" so that these rules no longer apply.

EVENT. In relativity theory this term refers to a point in four-dimensional spacetime, that is, an event is specified by its location in space (three dimensions) and the time at which it takes place (the fourth dimension). Events are located by world points in a spacetime or Minkowski diagram. See our discussion in section 3.11.

FIELD. One can think of forces as acting from one object directly on another (for example, the force of gravity as a force produced by one mass reaching out across empty space and pulling directly on another mass) or as due to a modification of space. This modification of space is called a "field." For example, one can say that the presence of matter causes a gravitational field; when a second piece of matter is present in the region of space within which this field exists, that second mass experiences a gravitational force. The field concept is much more useful to physicists than the direct-force or "action-at-a-distance" model in that it permits the creation of physical models with a much wider domain of validity. See our discussion of electric and magnetic fields in section D.2.

FLAT space (Euclidean space). When no matter is present, according to general relativity, the rules of Euclidean geometry should obtain in four-dimensional spacetime, and in particular geodesics (the shortest path between two points) will be straight lines. Such a region of space is said to be "flat" or "Euclidean" to distinguish it from "curved" or "warped" or "non-Euclidean" regions of space where the rules of Euclidean geometry do not hold. The terms "Euclidean" and "non-

Euclidean'' are much preferable since ''curved space'' or ''flat space'' may suggest arcane or even mystic matters. We have tried to avoid such terms in this book, but they will be found elsewhere.

FREQUENCY. In speaking of wave phenomena one characterizes the wave by three numbers: the speed at which the wave crests move through space (called the wave ''speed''); the distance in space separating successive wave crests (called the ''wavelength''); the number of crests passing any point in space each second (called ''the frequency''). The frequency or wavelength of a light wave determine the color of the light. The product of the frequency and the wavelength is equal to the speed of a given wave.

GAMMA RAYS. Electromagnetic waves having extremely short wavelengths (or high frequency) are called gamma rays. They are emitted in nuclear reactions on earth, and are received from various sources in space by gamma ray telescopes operating above the earth's atmosphere.

GENERAL RELATIVITY. Einstein's relativity theory of 1915 was called ''general'' because, unlike his 1905 relativity theory that holds only for inertial observers, the ''general theory'' of relativity holds for observers in all states of motion.

GEODESICS. In geometry these are lines representing the shortest distance between two points in space. The geometry can be characterized by any number of dimensions. Relativity treats geodesics in four-dimensional spacetime. If spacetime is ''flat'' or ''Euclidean,'' then the ordinary rules of schoolbook geometry apply and, in particular, a geodesic (the shortest distance between two points) is a straight line. Geodesics are not straight in non-Euclidean spaces. See the discussion in section 5.6.

GRAVITATION. In Newtonian terms, any piece of matter exerts a force of attraction on every other piece of matter in the universe. This attractive force is called the force of gravity. It may be thought of as due to one piece of matter acting directly on another (so-called action-at-a-distance); or, more fruitfully for physicists, a piece of matter may be thought of as establishing a modification of the surrounding space, a modification characterized by a gravitational field. Other pieces of matter moving in the region of space in which this field exists will experience a gravitational force. According to Einstein's general theory of relativity, what we term gravitation is really a manifestation of the departure of spacetime from Euclidean geometry in a given region of space. Thus objects move nonuniformly in such regions of space not because a gravitational force acts upon them but because their geodesics in spacetime are non-Euclidean. See section 5.3.

INERTIA. This property of any object describes the reluctance the object has to any change in its state of motion (its speed or its direction of travel). Inertia is given quantitative expression as a number representing the mass of the object. The most common units of mass are the gram or kilogram (and very seldom in the United States a unit called the "pound" is used, although this should not be confused with the unit of *force* called the pound). See section 2.5.

INERTIAL REFERENCE SYSTEM. This is a reference frame in which Newton's model of motion (in particular his first law) holds. Any other reference frame moving uniformly (that is, with no acceleration—at constant speed and in a constant direction) with respect to an inertial frame is itself an inertial reference frame. See the discussion in section 2.4.

INFRA-red rays or radiation. This is a form of electromagnetic radiation like light but of much lower frequency or much longer wavelength.

INTERVAL. In special relativity theory this quantity is an invariant for all inertial observers (as the speed of light is postulated to be). A formula for the interval in special relativity is given in section A.4.

LIGHT. A form of electromagnetic radiation to which the human eye is sensitive.

LORENTZ TRANSFORMATIONS. These equations summarize the results of Einstein's special (1905) theory of relativity. They provide algebraic procedures for determining the time and distance measurements of events as measured by any inertial observer in terms of measurements of the same event made by any other inertial observer. They are named for the physicist H. A. Lorentz who published the same equations some years before Einstein's work, although Lorentz did not recognize their significance at that time. See our discussion in section 2.7, and appendixes A, B, and C.

LUMINIFEROUS ETHER. This was thought to be the medium pervading all of space that carried light waves. There was never substantial evidence for the existence of the ether except for the fact that physicists could not imagine wave motion (a characteristic that light was known to display) without some medium. Hence its existence was postulated. It is a concept no longer used in physics. See sections D.4 and 2.7.

LUNARY SPHERE. In the Ptolemaic model of the cosmos, each of the celestial bodies was attached to a sphere centered on the earth. The sphere containing the moon, the lunary sphere, was of special significance because within this sphere all matter was thought to be mutable, whereas beyond this sphere only the divine and immutable ether existed.

MAGNETISM. Certain materials (loadstone, for example) show a natural capability of modifying the properties of space around them: they are capable of setting up a "magnetic field" that can act on other objects susceptible to magnetic forces. Magnetism is one way of describing this capability. Magnetic forces also act on electrically charged objects, and Maxwell's equations show that a magnetic field that changes in time will produce an electric field; conversely, an electric field that changes in time will produce a magnetic field. See appendix D for a complete discussion.

MASS, inertial and gravitational. Inertial mass is the quantitative measure of inertia, or the reluctance a body has to undergo a change in its motion. See our discussion in section 2.5. Gravitational mass is a concept distinct from inertial mass. Gravitational mass is a measure of the degree to which a body responds to the presence of a gravitational field (or, equivalently, it measures how strongly the body acts on and is acted upon by other masses through the gravitational force). See our discussion in section 5.3. Both inertial and gravitational masses are measured in the same units (generally grams or kilograms) and are numerically equal.

MAXWELL'S EQUATIONS. These equations provide a comprehensive mathematical description of observed electric and magnetic phenomena. The experimental work of Michael Faraday was of special importance to Maxwell in his formulation of these equations. See our discussion in section 2.7 and in appendix D.

MECHANICS. This branch of physics deals with forces and the way that objects respond to forces. The basis of "classical mechanics" (that is, without relativity or quantum phenomena) is Newton's *Principia*.

MESON. This is one of the "subatomic" or "elementary" particles. Several types of mesons are known. Their significance for the discussion in this book is that they have a limited and well-determined average lifetime. They can therefore be used as a clock. When they move at high speeds relative to an observer, this clock "runs slow," that is, the mesons take longer to decay than when they are at rest. This provides one experimental confirmation of the phenomenon of time dilation predicted in the special theory of relativity. See section 3.9.

MOMENTUM. Newton called this "the quantity of motion" in the *Principia*. Momentum is defined as the mass of a body multiplied by its speed. The direction of the motion must also be specified. See section 2.5.

MU-MESON. One of the sorts of mesons. See MESON.

NEUTRAL PI-MESON. One of the meson particles. See MESON.

NEUTRINO. This is one of the "subatomic" or "elementary" particles found in nature. The neutrino is generally believed to have no mass and to move at the speed of light. (This gloss is qualified somewhat because there is recent [1985] and disputed evidence that the neutrino might have a small mass. In this case, according to relativity theory, it will not move at the speed of light. Research is currently underway to settle the point.)

OPTICS. Optics is the branch of physics dealing with light, and more broadly, with any sort of electromagnetic radiation. Thus one can speak of x-ray optics, or of microwave (radio) optics.

PARALLEL. This term applies to geodesics that do not cross. In our usual Euclidean geometry, geodesics (the shortest path between two points in space) are straight lines, and so parallel lines are straight lines that never cross. In non-Euclidean geometries geodesics are not straight lines, but the adjective "parallel" may still be applied if two geodesics never cross.

PERIOD. In any oscillatory or cyclic phenomenon the "period" is the time between consecutive cycles. Thus in a pendulum clock, the period is the amount of time between swings of the pendulum, or in the case of the earth's cyclic motion around the sun, the "period" of the earth is the time for one complete cycle about the sun (one year).

PRIMUM MOBILE (or "prime mover"). This was the outermost layer of the Ptolemaic cosmos, and it was responsible for the motion of all of the other, inner spheres. See figure 1.1.

PTOLEMAIC MODEL of the cosmos. This refers to the model created by Ptolemy, which became the dominant cosmological model in Europe in the Middle Ages. The term is also used as an equivalent to the generic term "geocentric model." See section 1.3.

PYTHAGOREAN THEOREM. This is the relation in Euclidean geometry between the lengths of the sides of a right triangle (a triangle containing one "right" or 90-degree angle). The square of the hypotenuse (the longest side of the triangle) is equal to the sum of the squares of the other two sides. In algebraic terms, if C is the length of the hypotenuse, and A and B are the lengths of the other two sides of the triangle, then: $C^2 = A^2 + B^2$.

QUANTUM. In modern (as opposed to classical) physics, certain entities once thought to be continuous are now considered to consist of discrete units called "quanta." The branch of modern physics called "quantum mechanics" works with models of the physical world based on this quantum concept. Examples of "quantized" entities once thought to be continuous include energy and momentum. The particle model of light regards each light particle as a quantum of electromagnetic energy (called a "photon").

RADIATION. As used in this book, this term refers to electromagnetic radiation.

RADIO WAVES. Radio waves constitute one form of electromagnetic radiation having a much longer wavelength (or a much lower frequency) than light.

RELATIVISTIC. This adjective is applied whenever the effects of Einstein's relativity theory become significant. Newton's models work very well when objects move past an observer much more slowly than the speed of light, and the difference between the results of Einstein's and Newton's analyses is nearly negligible. Such situations are termed "nonrelativistic." When things move at speeds comparable to a few tenths of the speed of light, relativity may give answers significantly different from those provided by Newton's models (depending on the accuracy desired in the results); if the difference is considered significant, the term "relativistic" is applied. The point at which the term is applicable is not rigorously defined and depends on both the physical circumstances and the accuracy of measurement desired.

RELATIVITY, principle of. In its restricted or "special" version, this principle was one of Einstein's two postulates in his 1905 special theory of relativity. It asserts that the laws of physics will be of the same form (the equations describing the laws will be invariant) for all inertial observers. See our discussion in section 3.6. In its later form in the general theory of relativity, Einstein asserted that the equations (laws) of physics must be invariant for all observers regardless of their state of motion. This general statement obviously includes the ''special'' principle of relativity as a special case. See section 5.2.

RETROGRADE. Objects in the heavens move from east to west in their daily (diurnal) cycle across the sky. The sun, moon, and planets, however, have an additional motion. They move slowly to the east with respect to the background stars. Planets sometimes reverse their direction of travel with respect to the stars and travel westward for a brief time and then resume their gradual eastward drift. This temporary ''backward'' movement is called ''retrograde'' motion of the planets, and it was one of the central problems faced by pre-Copernican scientists attempting to create accurate models of the heavens.

RIGID. A rigid body is one whose various parts do not shift in relation to one another. No such object has ever been encountered. All actual matter has some flexibility and will respond to forces in a nonrigid way. Nevertheless, the idealization is useful in analyzing mechanical situations and for many purposes can be closely approximated by objects at hand.

SPACETIME. This is the term given to the four dimensions used to describe events in relativity theory. It is often represented in a ''spacetime diagram'' or ''Minkowski diagram.'' It is important to bear in mind that although space and time are treated in a similar fashion in these diagrams, they are not at all the same, nor do space and time become ''confused'' with one another or ''mixed-up'' or even ''interchanged,'' as is sometimes asserted in discussions of relativity theory. See sections 3.11 and 5.7.

SPECIAL RELATIVITY. This refers to Einstein's original relativity theory published in 1905. The theory is ''special'' in that it treats situations involving inertial (unaccelerated) observers only. His later ''general theory'' had a domain of validity that covered accelerating observers and included ''special relativity'' as a special case.

SPECTRUM (or ''electromagnetic spectrum''). This is a way of presenting all of the different forms of electromagnetic energy. See the discussion in section 2.7 and especially box 2.1.

SPEED. When an object is moving with respect to an observer, the speed of the object is defined to be the distance traveled by the object divided by the time required to travel that distance. A related term is VELOCITY, which, like speed, is a number found from dividing the distance traveled by the time required to travel that distance; but velocity also involves a specification of the *direction* of motion. Thus velocity requires two things to be specified (speed and direction), whereas speed requires only one number. (Quantities like velocity that have a number and a direction in space associated with them are called "vectors.")

TACHYONS. Tachyons are hypothetical particles that go faster than light. See appendix C.

TENSORS. These mathematical constructs describe conditions at each point in spacetime (actually, tensors may be defined for spaces having any number of dimensions). For example, the matter, energy, electric, or magnetic fields present at each point in spacetime may be described with a tensor. Because the tensor refers to some property intrinsic to each point in space, it is independent of any reference frame. Hence tensor equations are invariant, and laws of physics expressed as tensor equations are invariant. Following the rules of tensor calculus, one may find the measured value of the quantity to which the tensor refers for any specified reference frame. See the discussion in section 5.9.

ULTRA-VIOLET. This is a form of electromagnetic radiation having a frequency much higher (or a wavelength much smaller) than light.

UNIFORM MOTION. This is unaccelerated motion (that is, motion in a straight line and with unchanging speed).

VELOCITY. This quantity is used to describe the motion of an object. It involves specification of both the speed and the direction of travel.

WORLD LINE, WORLD POINT. These terms refer to the location of events in a spacetime diagram (or Minkowski diagram). Points in these diagrams refer to events; they are also called "world points." Lines in the diagrams refer to paths followed by moving entities in spacetime (for example, Gertrude's train, a flash of light, or a planet orbiting the sun). These lines are called "world lines" of the moving entity. See our discussions in sections 3.11 and 5.7.

Suggestions for Further Reading

(A more extensive annotated list may be gleaned from the footnotes.)

Bernstein, Jeremy. *Einstein*. New York: Penguin Books, 1973. This is a brief but extremely well-written intellectual biography of Einstein for general readers.

Clark, Ronald. *Einstein: The Life and Times*. New York: Avon Books, 1971. Clark's biography is much longer and more thorough than Bernstein's; it is fully documented and includes an extensive bibliography. It is intended for general readers.

Einstein, Albert. *Relativity: the Special and General Theory*. New York: Crown Publishers, 1961. This is Einstein's own attempt at "popularizing" his theories of relativity. Anyone who has completed the present book should have no trouble with this explanation of Einstein's.

Einstein, Albert, and Leopold Infeld. *The Evolution of Physics*. New York: Simon and Schuster, 1966. Einstein and Infeld wrote this splendid book for general readers to explain the process and the results of certain areas of modern physics. Without some background in physics, general readers may find *The Evolution of Physics* challenging or even a bit perplexing at points.

Ferris, Timothy. *The Red Limit*. New York: Quill, 1983. Ferris's book is an outstanding piece of science writing for general readers. He explains key theoretical and observational advances made in the ongoing quest for solutions to the problems of cosmology.

Holton, Gerald, and Yehuda Elkana, eds. *Albert Einstein: Historical and Cultural Perspectives*. Princeton, N.J.: Princeton University Press, 1982. This book consists of papers delivered at a symposium honoring the hundredth anniversary of Einstein's birth. Most of the sections require some specialized knowledge on the part of the reader.

Kaufmann, W. J. *The Cosmic Frontiers of General Relativity*. Boston: Little, Brown and Company, 1977. This is appropriate as a college-level text for nonscience students on general relativity and cosmology. It includes discussions of black holes and relativistic space-

flight. Readers of the present book should have little difficulty with
Kaufmann's explanations.

Pais, Abraham. *Subtle Is the Lord: The Science and the Life of Albert
Einstein*. New York: Oxford University Press, 1982. This is probably the definitive scientific biography of Einstein. Pais is writing for
an audience with a background in physics and mathematics through
tensor calculus, but some sections of the book are readily accessible
to general readers.

INDEX

Library of Congress Cataloging-in-Publication Data

Mook, Delo E., 1942–
    Inside relativity.

    Bibliography: p.
    Includes index.
    1. Relativity (Physics) I. Vargish, Thomas.
II. Title.
QC173.55.M66 1987   530.1′1   87–45528
ISBN 0–691–08472–6 (alk. paper)